Master Math: Geometry

By

Debra Anne Ross

Course Technology PTR

A part of Cengage Learning

COURSE TECHNOLOGY
CENGAGE Learning

Australia • Brazil • Japan • Korea • Mexico • Singapore • Spain • United Kingdom • United States

COURSE TECHNOLOGY
CENGAGE Learning

Publisher and General Manager,
Course Technology PTR:
Stacy L. Hiquet

Associate Director of Marketing:
Sarah Panella

Manager of Editorial Services:
Heather Talbot

Marketing Manager: Jordan Casey

Senior Acquisitions Editor:
Emi Smith

Interior Layout Tech:
Judith Littlefield

Illustrations and Equations:
Judith Littlefield and
Mike Tanamachi

Cover Designer: Jeff Cooper

Indexer: Larry Sweazy

Proofreader: Jenny Davidson

For product information and technology assistance, contact us at
Cengage Learning Customer & Sales Support, 1-800-354-9706
For permission to use material from this text or product,
submit all requests online at **cengage.com/permissions**
Further permissions questions can be emailed to
permissionrequest@cengage.com

All trademarks are the property of their respective owners.

Library of Congress Control Number: 2009924623

ISBN-10: 1-59863-984-6
ISBN-13: 978-1-59863-984-1

Course Technology, a part of Cengage Learning
20 Channel Center Street
Boston, MA 02210
USA

Cengage Learning is a leading provider of customized learning solutions with office locations around the globe, including Singapore, the United Kingdom, Australia, Mexico, Brazil, and Japan. Locate your local office at:
international.cengage.com/region

Cengage Learning products are represented in Canada by Nelson Education, Ltd. For your lifelong learning solutions, visit **courseptr.com**
Visit our corporate website at **cengage.com**

Printed in Canada
1 2 3 4 5 6 7 12 11 10 09

Table of Contents

Acknowledgments

I would sincerely like to thank Kristin Peterson for her careful reading of this book and for all her invaluable comments and suggestions. I am also indebted to Dr. Melanie McNeil, Professor of Chemical Engineering at San Jose State University, for her careful reading of this book and for all her valuable comments. I am grateful to Dr. Channing Robertson, Professor of Chemical Engineering at Stanford University, for reviewing this book and, in general, for all his guidance. I especially thank my mother, Maggie Ross, for her editorial help.

Without my wonderful agent, Sidney B. Kramer, and the staff of Mews Books, the *Master Math* series would not have been published. Thank you, Sidney! I am also thankful to Ron Fry and the staff of Career Press for their work in publishing and launching the original *Master Math* books as a successful series.

I am grateful to Emi Smith, Senior Acquisitions Editor, and Course Technology, a part of Cengage Learning, for invigorating the *Master Math* series and improving the presentation. I particularly appreciate Judith Littlefield for her tireless and expert work on the illustrations, equations, and layout. I also thank Mike Tanamachi for his work on illustrations. Much thanks to Jenny Davidson for proofreading, Jeff Cooper for cover design, Larry Sweazy for indexing, as well as Stacy L. Hiquet, Sarah Panella, Heather Talbot, and Jordan Casey.

Finally, I deeply appreciate my beautiful and brilliant husband, David A. Lawrence, who worked side-by-side with me as we meticulously edited text and figures.

About the Author

Debra Anne Ross Lawrence is the author of six books of the *Master Math* series: *Basic Math and Pre-Algebra, Algebra, Pre-Calculus, Calculus, Trigonometry*, and *Geometry*. She earned a double Bachelor of Arts degree in biology and chemistry with honors from the University of California at Santa Cruz and a Master of Science degree in chemical engineering from Stanford University.

Her research experience encompasses investigating the photosynthetic light reactions using a dye laser, studying the eye lens of diabetic patients, creating a computer simulation program of physiological responses to sensory and chemical disturbances, genetically engineering bacteria cells for over-expression of a protein, and designing and fabricating biological reactors for in-vivo study of microbial metabolism using nuclear magnetic resonance spectroscopy.

Debra was a member of a small team of scientists and engineers who developed and brought to market the first commercial biosensor system. She managed an engineering group responsible for scale-up of combinatorial synthesis for pharmaceutical development. She also managed intellectual property for a scientific research and development company. Debra's work has been published in scientific journals and/or patented.

Debra is also the author of *The 3:00 PM Secret: Live Slim and Strong Live Your Dreams* and *The 3:00 PM Secret 10-Day Dream Diet*. She is the coauthor with her husband, David A. Lawrence, of *Arrows Through Time: A Time Travel Tale of Adventure, Courage, and Faith*. Debra is President of GlacierDog Publishing and Founder of GlacierDog.com. When Debra is not engaged in all-season mountaineering near her Alaska home, she is endeavoring to understand the incomprehensible workings of the universe.

Introduction

Geometry is present in nature, art, architecture, surveying, navigation, cartography, biology, chemistry, physics, geology, astronomy, all fields of engineering, and in the structure of the smallest bits of matter to the grandest galaxies. The study of geometry is like detective work—you are given bits of information and use logic and reasoning to determine what you want to know. Becoming proficient in geometry will train your mind to solve problems in a creative and efficient way—like a detective.

In forming the field of geometry, ancient mathematicians developed the *postulation system* in which one begins with a set of unproved statements or postulates, and deduces using logic, other statements, or theorems. Accordingly, the development of logic and deductive reasoning was instituted to prove geometric statements. Geometry was used by ancient people including Babylonians, Egyptians, Romans, and Greeks in practical applications such as land measurement, surveying, construction, navigation, and astronomy. Information and facts pertaining to geometry were organized and developed by Greeks between 600 and 300 B.C., and described by Euclid in his famous book *Elements* in approximately 300 B.C. *Euclidean geometry* combines related elements using the methods of logic and reasoning, and the tools of axioms, postulates, definitions, theorems, and constructions in order to prove, describe, calculate, generate, or use information pertaining to geometric objects. Euclid provided five primary postulates which can be described as: (1) One straight line connects any two points; (2) Any straight line can be extended infinitely in either direction; (3) A circle can be drawn with any center and any radius; (4) All right angles are equal; and (5) Given a line and a point not on the line, only one line can be drawn parallel to that line through the point. The process by which mathematicians attempted to verify this fifth *parallel line postulate* led to non-Euclidean geometries.

Euclidean geometry describes the world we think we see around us in which the shortest distance between two points is a straight line, the angles in a triangle always sum to 180°, and parallel lines lie in the same plane, remain equidistant, and never intersect even if they are *infinitely* long. *Non-Euclidean geometries* are less obvious. In *spherical geometry*, which takes place on a sphere and is used by pilots, ship's captains, and astronomers, no parallel lines exist, the angles in a triangle sum to greater than 180°, and the shortest distance between two points is a great circle (the largest circle that can be drawn through any point on a sphere). In *hyperbolic geometry*, which can be represented in two dimensions as saddle-shaped, the angles in a triangle sum to less than 180°, and through a point not on a line, there is *more than one* line parallel to that line. Euclidean geometry provides an excellent representation for part of the universe that we observe, but in the study of certain aspects of our universe, or the universe itself, non-Euclidean geometries may provide a more accurate portrayal. For example, in Einstein's Theory of General Relativity, matter produces curved space-time.

Euclidean triangle   Hyperbolic triangle   Spherical triangle

Another branch of geometry developed in the 17th century by René Descartes is *coordinate geometry,* also called *analytic geometry*, which is the study of geometry using the analytical methods of algebra. This approach involves placing a geometric figure into a coordinate system illustrating a proof, and obtaining information about the figure using algebraic equations.

Today, geometry has been joined with computers and computer-aided design and is used in fields such as automobile manufacturing, computer vision, robotics, video game programming, virtual reality, aerospace, and architecture. Architecture examples include the innovative work of Frank O. Gehry in his Guggenheim Museum in Bilbao, Spain, and Norman Foster's striking glass and steel London City Hall.

Master Math: Geometry provides everything a high school or first year college student needs to know including an explanation of deductive reasoning, how to perform proofs and constructions, as well as definitions, theorems, postulates, and examples pertaining to points, lines, planes, angles, ratios, proportions, triangles, congruence, similarity, quadrilaterals, polygons, circles, surface area and volume of geometric solids, and coordinate, or analytic, geometry. *Master Math: Geometry* is part of the *Master Math* series, which includes *Master Math: Basic Math and Pre-Algebra, Master Math: Algebra, Master Math: Trigonometry, Master Math: Pre-Calculus, Master Math: Solving Word Problems*, and *Master Math: Calculus.* This book and those previously listed are written to provide clear, easy to understand, comprehensive reference sources that allow quick access to explanations of concepts, principles, definitions, examples, and applications. *Master Math: Geometry* is written to assist high school and college students, teachers, tutors, and parents, as well as to serve as a reference for scientists, engineers, architects, or anyone needing a basic reference.

Chapter

1

The Language of Geometry, Deductive Reasoning and Proofs, and Key Axioms and Postulates

This chapter provides a mini-reference to the book by defining words, symbols, and terms used in geometry so that you can familiarize yourself with the language of geometry. Axioms and postulates used in basic geometry and throughout the book are also listed. Axioms and postulates are statements pertaining to mathematics and geometry, respectively, that are assumed to be true. Deductive reasoning and *how to* write a proof are also explained. Note also that topics presented in this book are often discussed in more than one section. The index can be used to locate relevant pages for each topic.

1.1 The Language of Geometry

• Geometry is made up of somewhat unfamiliar terms, such as construction, axiom, postulate, theorem, corollary, congruence, similarity, circumscribed, inscribed, Euclidean, non-Euclidean, hyperbolic, spherical, and deductive reasoning, as well as the more familiar words we would expect in geometry, such as point, line, plane, angle, triangle, circle, polygon, coordinate system, square, area, length, volume, surface area, ratio, and proportion. Not only does geometry have a language all its own, but while we may initially think of it as purely a 'math' subject, it also involves logical reasoning and proofs. The practice of proving a geometric theorem is not only foundational to geometry, but also to the development of our thinking and reasoning skills. Two of the most important aspects of geometry are gaining expertise in logical reasoning, and learning how to calculate the various features of geometric figures, which are often models of physical structures in the world around us.

Symbols and Terms and Their Meanings:

m	*measure of,* as in $m\angle A$.
$\angle$	*angle,* as in $\angle A$.
∟	The box denotes a *90 degree,* or *right,* angle.
$\overleftrightarrow{AB}$, $\overline{AB}$, $\overrightarrow{AB}$	*line AB, segment AB, ray AB,* respectively. *Lines* and *segments* are equivalently written using italics as *AB*.
$\cong$	*is congruent to.*
~	*is similar to.*
‖	*is parallel to.*
⊥	*is perpendicular to.*
°	*degrees.*
Δ	*triangle.*
$\sqrt{x}$	*square root of x,* also written $x^{1/2}$.
$\sqrt[3]{x}$	*cubed root of x,* also written $x^{1/3}$.
$\sqrt[n]{x}$	*nth root of x,* also written $x^{1/n}$.
Axiom:	A statement pertaining to the field of mathematics that is accepted as true. Also called *algebraic postulate*.
Circumscribed/Inscribed:	A polygon is said to be *circumscribed* about (around) a circle if each side of the polygon is tangent to the circle. The polygon is outside the circle, and the circle is *inscribed* inside the polygon. See (a) following.

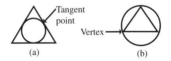

Tangent point

Vertex

(a)   (b)

A circle is said to be *circumscribed* about (around) a polygon if each vertex of the polygon is on the circle. The circle is outside the polygon, and the polygon is *inscribed* inside the circle. See (b) in preceding figure.

Congruence: Congruent figures are figures having the same size and shape. Congruent angles have equal measures. Congruent circles have equal radii. Congruent segments have equal lengths. Congruent polygons have all corresponding sides and angles of equal measure. Congruent figures are images of each other and are said to have isometry. Congruent objects are the same, but may be in a different location.

Constructions: Constructions are geometric drawings made with a straightedge and a compass. A straightedge is used to draw lines and a compass is used to draw circles and arcs. Arcs are used to locate points and points are used to determine lines. (See Chapter 8.)

Coordinate geometry, also called *analytic geometry*: Combines Euclidean geometry with algebra. It is the study of geometry using the analytical methods of algebra. This approach involves placing a geometric figure into a coordinate system and obtaining information about the figure using algebraic equations. (See Chapter 9.)

Corollary: A theorem that is directly related to another theorem and can be easily proved because of that theorem.

Deductive reasoning: Proving a statement or theorem in a logical manner using axioms, postulates, definitions, and proved theorems. Deductive reasoning is used in formal proofs and begins with general assumptions and truths and derives specific conclusions. (See Section 1.2.)

Direct proof: A method of proof in which premises lead to a conclusion using deductive reasoning and already accepted statements. A direct proof is the most common method used to prove theorems. (See Section 1.3.)

Euclidean geometry: Euclidean geometry was described by Euclid in his famous book *Elements* in about 300 B.C. *Euclidean geometry* combines related elements using the methods of logic and reasoning and the tools of axioms, postulates, definitions, theorems, and constructions in order to prove, describe, calculate, generate, or use information pertaining to geometric objects. In Euclidean geometry, the *parallel line postulate* holds true: *Through a given point not on a line, there is one and only one line parallel to that line.* Parallel lines lie in the same plane and never intersect in Euclidean geometry, even when they are *infinitely long*. In Euclidean geometry, the angles in a triangle sum to 180°.

Hyperbolic geometry: A non-Euclidean geometry that when represented in two dimensions is saddle-shaped. In hyperbolic geometry the parallel postulate fails so that: *Through a given point not on a line, there is MORE THAN ONE line parallel to that line.* Also, in hyperbolic geometry, the angles in a triangle sum to *less than* 180°.

Hypothesis and conclusion: Pertains to conditional statements in the
if-then form or in the *subject-predicate* form. For example,
in the statement:
If the polygon has three sides, *then* it is a triangle, the
phrase "*If* the polygon has three sides" is the hypothesis,
and "*then* it is a triangle" is the conclusion.

The *subject-predicate* form is:
A polygon with three sides is a triangle, where "A polygon
with three sides" is the hypothesis/subject, and "a triangle"
is the conclusion/predicate. Theorems can be written in
either form of a hypothesis/conclusion conditional statement. (See Section 1.3.)

Indirect proof: An indirect proof involves beginning with the negative
or opposite of the desired proof and reasoning through
the proof until reaching a contradiction between one of
the statements and the negative of the desired proof. The
contradiction occurs when one of the statements asserts
that the negative of the desired conclusion cannot be true
and therefore the desired conclusion must be true. (See
Section 1.3.)

Inductive reasoning: In inductive reasoning, a conclusion is based on or
drawn from past observations. Inductive reasoning begins
with specific assumptions and truths and concludes
something general, which may or may not always hold
true in all cases. For example, because we see most birds
fly, we may conclude that all birds fly, but this is not
always true. (See end of Section 1.2.)

Isometry: A transformation that preserves lengths and distances,
and maps each segment to a congruent segment.

Locus: The locus is the set of points that satisfies a certain
condition or conditions. For example, a perpendicular
bisector represents the locus of points equidistant
between two points on the line connecting the points. In
coordinate, or analytic, geometry, the locus is represented
by a graph of an equation as the points or coordinates
that satisfy the equation. (See Sections 8.6 and 9.6.)

Non-Euclidean geometry: In non-Euclidean geometry, the parallel line
postulate (*through a given point not on a line, there is
one and only one line parallel to that line*) does not hold
true. Non-Euclidean geometries include hyperbolic and
spherical/elliptic.

Plane geometry: Two-dimensional Euclidean geometry is called plane
geometry. Plane geometry studies planar figures that can
be constructed using a straightedge and compass.

Postulate: A statement pertaining to geometry that is assumed to be true without proof.

Proofs: A proof involves logical reasoning using definitions, axioms, postulates, and theorems, all accepted as true in order to prove a statement or theorem. A formal proof includes a statement or theorem, what is *given* (the *hypothesis*), what is to be *proved* (the *conclusion*), a figure, and a two-column listing comprising a logical flow of statements used to build the proof on the left and the reasons why each of these statements are valid on the right. *Statements* include information given, axioms, postulates, definitions, and proved theorems. (See Section 1.3.)

Similar polygons: Similar polygons have the same shape, but not necessarily the same size. Similar polygons have corresponding angles congruent and corresponding sides in proportion.

Solid geometry: Three-dimensional Euclidean geometry. (See Chapter 7.)

Spherical/elliptic geometry: Non-Euclidean geometry that takes place on a sphere in which the parallel postulate fails so that *through a given point not on a line, there ARE NO lines parallel to that line*. Lines drawn on a sphere intersect, and no parallel lines exist. In spherical geometry, the angles in a triangle sum to *greater than* 180°.

Theorem: A statement that can be proven using definitions, axioms, postulates, and other proved theorems. Theorems in geometry are generally proved using deductive reasoning. Theorems are comprised of two parts—the *given* or *hypothesis*, which is what is known, and the *conclusion* or *proof*, which is what needs to be proven. Theorems can be written in the form of either *if-then* or *subject-predicate* conditional statements.

Transformations: A one-to-one correspondence between two sets of points or a mapping from one plane to another.

1.2 Deductive Reasoning

• One of the most important reasons to learn and understand geometry is to develop your ability to engage in logical thinking and reasoning. In many areas of our lives, we can excel if we apply reasoning and think through ideas, situations, or problems in a logical manner. Geometry has, in its foundation, the concept of logical reasoning and proving principles to be true by working through **proofs**.

• A proof generally involves applying ***deductive reasoning*** to prove a geometric statement or *theorem*. The purpose of a proof is to prove that a geometric statement or theorem is true. A ***theorem*** is a statement that must be proven because it is not automatically assumed to be true. A theorem can be proven using the process of deductive reasoning and building to the final proof statement using *definitions, axioms, postulates,* and other *proved theorems*, which are all accepted as true. An example of a theorem is: The sum of the measures of the angles in a planar triangle is 180°.

• Statements automatically accepted as true include *definitions, axioms,* and *postulates*. ***Axioms*** are statements pertaining to the field of mathematics that are accepted as true. ***Postulates*** are statements pertaining to geometry that are accepted as true. (Be aware that axioms are often referred to as postulates or algebraic postulates, and in many geometry books, the word *principle* will be used to refer to axioms, postulates, and/or theorems. In addition, certain theorems may be referred to as postulates and vice versa.)

• Before discussing the mechanics of performing a proof, we first need to understand deductive reasoning. ***Deductive reasoning*** is a process used to work through *proofs of theorems* and *geometric statements* and begins with *general assumptions* and truths and derives *specific conclusions*.

• The process of deductive reasoning can be structured in the form of a ***syllogism***, which is composed of:

• A *general statement* called the *major premise*;

• A *particular statement* called a *minor premise*; and

• A *conclusion* or deduction (which is the logical conclusion of applying the general statement to the particular statement).

If the first and second premises are true, then the conclusion will be true. An illustration often used to visualize a deductive reasoning syllogism is circles inside circles. Consider the following examples:

• **Example**: Deductive reasoning syllogism.

   Major premise: All birds have <u>wings</u>.

   Minor premise: All Blue Jays are <u>birds</u>.

   Conclusion: All <u>Blue Jays</u> have wings.

(Note that the Apteryx, a bird of New Zealand, has only rudiments of wings.)

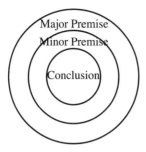 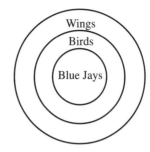

What if this example said:

Major premise: All birds have wings.

Minor premise: All Blue Jays have wings.

Conclusion: All Blue Jays are birds.

Because Blue Jays are a well-known bird, we know this is true. However, this example doesn't prove that Blue Jays are birds because a *Blue Jay*, which is only defined here as having wings, could also be an insect, which can have wings.

• **Example:** Deductive reasoning syllogism for supplementary angles, which are any two angles whose measures sum to 180°.

Major premise: Supplementary angles sum to <u>180 degrees</u>.

Minor premise: Angle A and Angle B are <u>supplementary angles</u>.

Conclusion: <u>Angle A and Angle B</u> sum to 180 degrees.

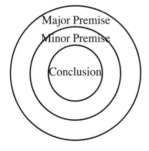 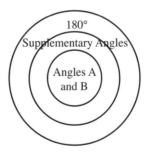

• The *major premise* of a syllogism can also be written as a ***conditional (if-then) statement***. For example, for the conditional statement, "If two angles are supplementary, then they sum to 180 degrees", a syllogism can be written:

If two angles are supplementary, then they sum to 180 degrees.

Angle A and Angle B are supplementary angles.

Angle A and Angle B sum to 180 degrees.

We can also rewrite the Blue Jay example in the if-then form as:

If it is a bird, then it has wings.

A Blue Jay is a bird.

Therefore, a Blue Jay has wings.

It would *not* be correct to say:

If it is a bird, then it has wings.

A Blue Jay has wings.

Therefore, a Blue Jay is a bird.

In this incorrect case, instead of supporting the *if* statement (If it is a bird), the second statement (A Blue Jay has wings) asserts the *then* statement (then it has wings), and the example is not logical.

Inductive Reasoning and Other Approaches to Proving a Truth

• *Inductive reasoning* begins with specific assumptions and truths and concludes something general, which may or may not always hold true in all cases. (Remember, deductive reasoning begins with general assumptions and truths and derives specific conclusions.) In inductive reasoning a conclusion is based on or drawn from past observations and while the conclusion may usually be true, there may be examples when it is not true. For example, we can cite numerous birds that fly and therefore conclude that all birds fly; however, there are examples of birds that don't fly such as the Ostrich, which has short wings, or the Apteryx, which has wing rudiments. Inductive reasoning can be valuable when attempting to understand a system or situation and is often used in observational sciences in order to gather facts and establish a theory or draw conclusions. It may be useful to make various observations and look for patterns, which can provide valuable information. When using inductive reasoning, it is, however, important to realize that there can be and often are exceptions.

• Proving something or establishing it to be true is also attempted by *observing*, *experimenting*, and *measuring*. These methods do not assure proof, but rather approximate values. Measurements have some degree of error, experiments are subject to the specific constraints designed into them by the experimenter, and observations can be subjective.

1.3 Theorems and How to Write a Proof

• A *theorem* is a statement that can be proven using definitions, axioms, postulates, and other proved theorems. Theorems in geometry are generally proved using deductive reasoning. Theorems are comprised of two parts— the *given* or *hypothesis*, which is what is known and the *conclusion* or *proof*, which is what needs to be proven. Theorems can be written in the form of either an ***if-then*** or a ***subject-predicate*** (declarative) conditional statement.

• ***Conditional statements*** have a hypothesis and conclusion and can be in the *if-then* form or in the *subject-predicate declarative* form as follows:

The *subject-predicate* form is a declarative statement.
For example: A cat has whiskers.

The *conditional (if-then) statement* has the form:
If x, then y, in which x is the ***hypothesis*** and y is the ***conclusion***.
For example: If it's a cat, then it has whiskers.

• Note: The ***if*** part of a conditional statement, which is the hypothesis, can be made up of one or more statements, and the ***then*** part forms the conclusion. Also, for a given *if-then* conditional statement, if there exists an example in which the *then* part is false while the *if* part is true, the whole statement is false.

• **Example:** Write a statement that a polygon is a three-sided triangle in both forms of a conditional statement.

Written in the *if-then* form we have:
"If the polygon has three sides, *then* it is a triangle"
where *"If* the polygon has three sides" is the *hypothesis* and *"then* it is a triangle" is the *conclusion*.

Written in the *subject-predicate* form this statement becomes:
"A polygon with three sides is a triangle"
where "A polygon with three sides" is the *hypothesis/subject* and "a triangle" is the *conclusion/predicate*.

• If we reverse the hypothesis and conclusion of a conditional *if-then* statement to *if y, then x,* the statement is called a ***converse of the conditional statement***. The converse of a conditional statement may or may not be true.

For example the conditional statement "if it is a cat, then it has whiskers" is true; however, the converse "if it has whiskers, then it is a cat" is not necessarily true because other animals also have whiskers.

An example where both conditional and converse are true is a definition:
If the angle measures 90°, then it is a right angle.
If it is a right angle, then it measures 90°.

This statement is a definition and the *converse of a definition is always true* because if x is what is being defined and y is the definition, then they will always be true regardless of the order. The definition of a right angle is that it measures 90°, and an angle measuring 90° is, by definition, a right angle.

- Other variations of the conditional statement *if x, then y* include:
(1) The ***inverse of a conditional statement***, which is the negative of both the hypothesis and conclusion, or *if not x, then not y*.
(2) The ***contrapositive of a conditional statement***, which is the negative of a conditional statement that has had the hypothesis and conclusion reversed, or *if not y, then not x*.

- A ***proof*** involves reasoning deductively using definitions, axioms, postulates, and proved theorems all accepted as true in order to prove a particular statement or theorem.

- The formal process of ***proving a theorem*** is:
(1) State the theorem and identify
     what is *given* (the *hypothesis*) and
     what is to be *proved* (the *conclusion*).
(2) Sketch a figure that illustrates the situation you are proving.
(3) Lay out a plan including statements and their respective axioms, postulates, definitions, and proved theorems you can use in your proof. (This is not technically part of a formal proof, but is useful.)
(4) Create a two-column listing comprising a logical flow of statements used to build the proof on the left, and the reasons why each of these statements are valid on the right. The last statement is what is to be proven.

- **Example:** Prove the theorem: The sum of the angles in a planar triangle is 180 degrees.

Given: $\triangle ABC$

Prove: $mLA + mLB + mLC = 180°$

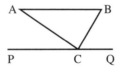

Plan: The plan is to draw a line (PQ) through one vertex point (C) that is parallel to the opposite side (AB) using the parallel line postulate. Then use the straight line PQ as a 180° angle and the parallel transversal postulate, which says that alternate interior angles of parallel lines are congruent.

Proof:

Statements	Reasons
1. Draw *PQ* ∥ *AB* through C of ΔABC.	1. Parallel line postulate: A line can be drawn parallel to a given line through an external point not on the line.
2. *mL*PCQ = 180°.	2. A straight line angle measures 180°.
3. *mL*PCA +*mL*ACB +*mL*BCQ = 180°.	3. Partition axiom: The whole equals the sum of its parts.
4. *L*A ≅ *L*PCA and *L*B ≅ *L*BCQ.	4. Parallel transversal postulate: Alternate interior angles of parallel lines (*PQ* ∥ *AB*) are congruent.
5. *mL*A +*mL*B +*mL*C = 180°.	5. Substitution axiom: A quantity may be substituted for an equal quantity. (#3, #4)

Therefore, we have proved the theorem: The sum of the angles in a planar triangle is always 180 degrees.

• **Example:** Prove the theorem: All right angles are congruent.

Given: *L*A and *L*B are right angles.

Prove: *mL*A = *mL*B, or equivalently, *L*A ≅ *L*B.

Plan: The plan is to use the definition stating that right angles equal 90 degrees and the transitive axiom, which states that things equal or congruent to the same or other things are equal or congruent to each other.

Proof:

Statements	Reasons
1. *L*A and *L*B are right angles.	1. Given.
2. *mL*A = 90° and *mL*B = 90°.	2. Definition: Right angles measure 90°.
3. *L*A ≅ *L*B.	3. Transitive axiom: Things congruent to the same are congruent to each other.

Therefore, we have proved the theorem: All right angles are congruent.

• Proofs are also written in an *informal* manner. An ***informal proof*** is an abbreviated form of a *formal proof*. There is the "Given", the "Prove", and the "Proof", which includes the main statements, such as "$\Delta ABC \cong \Delta DEF$ using SSS triangle congruence postulate" (without stating each sub-step between the main steps).

Direct Versus Indirect Proofs

• ***Theorems*** are generally written and proven using deductive reasoning and already accepted statements as ***direct proofs***. In a direct proof, there are premises that lead to a conclusion. In certain situations, a direct proof may be difficult, and an indirect proof may be used. *An **indirect proof** involves using a negative to establish something as being true.* Two possible results or conclusions are considered—the proof or desired conclusion and the negative of that proof or conclusion.

In working an ***indirect proof***, we begin with the opposite or negative of the desired conclusion and continue the process of the proof until reaching a contradiction between one of the statements and the negative of the desired proof. The contradiction establishes that the negative of the desired conclusion cannot be true, and therefore the desired conclusion must be true.

• **Example:** Use an indirect proof to prove that: If two angles are *not* congruent, then they are *not* both right angles.

Given: $\angle$A and $\angle$B are not congruent.

Prove: $\angle$A and $\angle$B are *not* both right angles.

We want to prove that $\angle$A and $\angle$B are *not* both right angles, so we first try to prove the opposite using an indirect proof:

$\angle$A and $\angle$B *are* both right angles.

Given: $\angle$A and $\angle$B are *not* congruent.

Prove: $\angle$A and $\angle$B *are* both right angles.

Given that $\angle$A and $\angle$B are not congruent, we attempt to prove (the negative) that $\angle$A and $\angle$B *are* both right angles:

All right angles are congruent was proved in the preceding proof.

If $\angle$A and $\angle$B *are* both right angles, then they must be congruent.

Except we are given that: $\angle$A and $\angle$B are *not* congruent.

This creates a contradiction and asserts that the negative of the desired conclusion cannot be true and therefore the desired conclusion must be true.

Therefore, since $\angle$A and $\angle$B are not congruent, then they are *not* both right angles.

1.4 Key Axioms and Postulates

• There are a number of important axioms and postulates that are frequently used in geometric proofs and problems and should be remembered. Axioms are statements pertaining to the field of mathematics that are accepted as true, and postulates are statements pertaining to geometry that are accepted as true. One source of possible confusion when learning geometry is that axioms are referred to as both axioms and postulates and certain postulates are referred to as theorems (or theorems are referred to as postulates). For example, Euclids's parallel postulate is called a theorem in certain books. Following are selected axioms and postulates with which you should be familiar. Each of these are discussed in context throughout the book.

Axioms, Also Called Algebraic Postulates

• *Transitive axiom*: **Things (quantities/values/objects) equal or congruent to the same or other things are equal or congruent to each other**.

　If $x = y$ and $y = z$, then $x = z$

　If $x = 3$ and $z = 3$, then $x = z$

• *Substitution axiom*: A quantity may be substituted for an equal quantity in any expression or equation.

　If $x = y$, then x can be substituted for y.

　$a + 6 = 10$ is equal to $a + (2)(3) = 10$

• *Partition axiom*: The whole equals the sum of its parts.

　$5 = 1 + 1 + 1 + 1 + 1$

　If $\overline{AB} = 3$, $\overline{BC} = 2$, $\overline{CD} = 2$

　then, $\overline{AD} = 3 + 2 + 2 = 7$

　(Note: Segments are also denoted as *AB*, *BC*, *CD*, and *AD*.)

• *Identity or reflexive axiom*: Any quantity is equal to itself.

　$64 = 64$

　$x + y = y + x$

　$\angle A = \angle A$, where $\angle$ is the *symbol for angle*.

　$\text{area}\triangle ABC = \text{area}\triangle ABC$, where $\triangle$ is the *symbol for triangle*.

- **Addition axiom:** If equals are added to equals, the sums are equal.

    $6 + 8 + 2$ is equal to $14 + 2$

    If $z = x + y$, then $z + z = (x + y) + (x + y)$

$$\begin{array}{r} (p = p) \\ + (q = q) \\ \hline p + q = p + q \end{array} \qquad \begin{array}{r} (4 = 4) \\ + (3 = 3) \\ \hline 4 + 3 = 4 + 3 = 7 \end{array}$$

- **Subtraction axiom:** If equals are subtracted from equals, the differences are equal.

    $(6 + 8) - 2$ is equal to $14 - 2$

    If $z = x + y$, then $z - z = (x + y) - (x + y)$

$$\begin{array}{r} (p = p) \\ - (q = q) \\ \hline p - q = p - q \end{array} \qquad \begin{array}{r} (4 = 4) \\ - (3 = 3) \\ \hline 4 - 3 = 4 - 3 = 1 \end{array}$$

- **Multiplication axiom:** If equals are multiplied with equals, the products are equal. Also, doubles of equals are equal.

    $(6 + 8) \cdot 2$ is equal to $14 \cdot 2$

    If $z = x + y$, then $z \cdot z = (x + y) \cdot (x + y)$

    If $x = 2$ and $y = 3$, then $x \cdot y = (2) \cdot (3)$

    If the area of a rectangle is 8, then the area of two rectangles is 16:

2	Area = 8	Area = 8	2	Total area = 16
	4	4		4 4

- **Division axiom:** If equals are divided by equals, the quotients are equal. Also, halves of equals are equal.

    $(6 + 8) / 2$ is equal to $14 / 2$

    If $z = x + y$, then $z/z = (x + y)/(x + y)$

    If $x = 2$ and $y = 3$, then $x/y = 2/3$

    If $m\angle A = m\angle B$, then $(1/2)m\angle A = (1/2)m\angle B$

    where m is the symbol for *measure of.*

- *Powers axiom*: Like powers of equals are equal.

  If $x = y$, then $x^2 = y^2$, $x^3 = y^3$, and $x^n = y^n$
  If $x = 2$, then $x^2 = 2^2 = 4$

- *Roots axiom*: Like roots of equals are equal.

  If $x = y$, then $x^{1/2} = y^{1/2}$, $x^{1/3} = y^{1/3}$, and $x^{1/n} = y^{1/n}$

Remember, the square root $\sqrt{x} = x^{1/2}$, the cubed root $\sqrt[3]{x} = x^{1/3}$, and the nth root $\sqrt[n]{x} = x^{1/n}$.

Geometric Postulates

- *One line postulate*: **One and only one straight line can be drawn between any two points.**

  Line *AB* is the only line that can be drawn between point A and point B.

- *Shortest distance postulate*: The shortest distance between any two points is the straight line that is drawn between them.

  The straight line is shorter than the bent line or the curve.

- *Line has two points postulate*: Any line contains at least two points.

- *Two points one line postulate*: Through any two points there is one and only one line.

- *Line intersection postulate*: Two straight lines can intersect each other at one and only one point.

- *Parallel (line) postulate*: **Through a given point not on a line, there is one and only one line parallel to that line.**

  Parallel line through point

  Line

- *Perpendicular through line postulate*: One and only one perpendicular line can be drawn to or through any point on a line (in a plane).

Point B on segment *AC* has one perpendicular line, *BD*, that is drawn to it.

- *Perpendicular point to line postulate*: One and only one perpendicular line can be drawn from or through any point *not* on a line to that line.

One line from external point D is drawn perpendicular to line *AC*.

- *Parallel transversal postulate*: **If two parallel lines are cut by a transversal, then corresponding angles are congruent**.

    A *transversal* is a line that intersects two other lines.

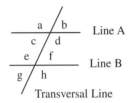

If lines A and B are parallel to each other, then **corresponding angles are congruent:**

$\angle a \cong \angle e$, $\angle b \cong \angle f$, $\angle c \cong \angle g$, and $\angle d \cong \angle h$.

Also:

Alternate interior angles are congruent: $\angle c \cong \angle f$ and $\angle d \cong \angle e$.

Vertical angles are congruent: $\angle a \cong \angle d$, $\angle b \cong \angle c$, $\angle e \cong \angle h$, $\angle f \cong \angle g$.

Adjacent angles (next to each other) are supplementary and sum to 180°.

- *Midpoint postulate*: Any straight line segment has one and only one midpoint.

Segment *AB* has M as its only midpoint.

Segments *AM* and *MB* have equal lengths.

- *Segment addition postulate*: If point B is between point A and point C, then: **segment *AB* + segment *BC* = segment *AC***

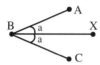

$$\overline{AB} + \overline{BC} = \overline{AC}, \text{ or } AB + BC = AC$$

- *Angle bisector postulate*: Any angle has one and only one bisector.

The bisector for $\angle$ABC is segment *BX*. The two angles formed by the bisector, $\angle$a and $\angle$a, are congruent.

- *Angle addition postulate*: If point B is in between point A and point C on the interior of angle AVC, then: $m\angle AVB + m\angle BVC = m\angle AVC$

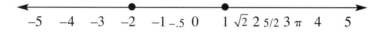

- *Ruler postulate*: The points on a line can be numbered so that:

  Every point corresponds to one and only one real number coordinate;
  Every real number coordinate corresponds to one and only one point;
  Every pair of points corresponds to one real number, which is the distance between the points; and
  The distance between two points is the absolute value of the difference between the coordinates of the points.

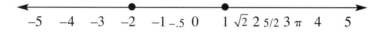

The two points depicted correspond to real numbers –2 and 1.
Real numbers –2 and 1 correspond to the two points depicted.
Points –2 and 1 have a *distance* between them that is the absolute
value of their difference: $|-2 - 1| = |1 - (-2)| = |-3| = 3$,
which is also the length of the segment between them. The ruler
postulate allows us to measure distance or determine length
between points on a line.

- **Protractor postulate**: The rays can be numbered so that: Every real
  number beginning with 0° and ending with 180° corresponds to one and
  only one ray; every real number coordinate corresponds to one and only
  one point; every pair of rays corresponds to one real number, which is
  the measure of the angle they determine; and the measure of the angle is
  the absolute value of the difference between the coordinates of its rays.
  (Note: A *ray* has an endpoint at one end and extends indefinitely in the
  other direction.)

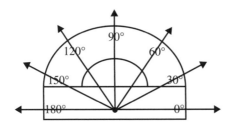

- **Arc sum postulate**: The measure of the arc formed by two adjacent arcs
  is the sum of the measures of the two arcs. In other words, if point B is
  on an arcABC, then the two arcs formed, arcAB and arcBC, sum to the
  total length of arcABC.

If point B is on arcABC, then
m arcABC $= m$ arcAB $+ m$ arcBC
where m is the *symbol for measure*.

- **Plane has three points postulate**: Any plane contains at least three noncollinear points. (Noncollinear means not in a line.)

Points are called *coplanar* if and only
if there is a plane that contains them.

- **Three points one plane postulate**: Through any three noncollinear
  points, there is one and only one plane. Also, through any three linear
  points, there is at least one plane.

Three noncollinear
points = one plane.

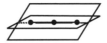

Three linear points =
more than one plane.

- ***Two points and line in plane postulate***: If two points lie in a plane, then the line joining them lies in the plane.

- ***Intersection of planes postulate***: If two planes intersect, their intersection is a line.

- ***Four points in space postulate***: Space contains at least four points not in the same plane.

- ***One circle per radius postulate***: One and only one circle can be drawn for a given radial distance r about any center point (in a plane).

    Only one circle can be drawn about the center point C for each radius r_1 and r_2.

- ***Change position postulate***: Any geometric figure can be moved or relocated to a new position without changing the figure's size or shape.

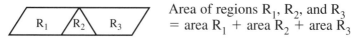

- ***Area addition postulate***: The area of an entire region is the sum of the areas of non-overlapping regions within it.

   Area of regions R_1, R_2, and R_3
= area R_1 + area R_2 + area R_3

- ***Area congruence postulate***: If two figures are congruent then they have equal areas.

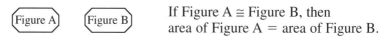   If Figure A $\cong$ Figure B, then area of Figure A = area of Figure B.

- *Area square/rectangle postulate*: The *area of a square* is the square of a side length, or A = s². The *area of a rectangle* is the length of its base multiplied by its altitude or height.

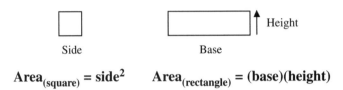

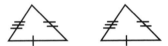

- *SSS triangle congruence postulate*: (Side-Side-Side) **If the three sides of one triangle are congruent to the three sides of another triangle, then the triangles are congruent.** (Tick marks indicate congruent side lengths.)

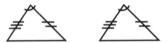

- *SAS triangle congruence postulate*: (Side-Angle-Side) **If two sides and the included angle of one triangle are congruent to two sides and the included angle of another triangle, then the triangles are congruent.** (Tick marks and arcs indicate congruent sides and angles.)

- *ASA triangle congruence postulate*: (Angle-Side-Angle) **If two angles and the included side of one triangle are congruent to two angles and the included side of another triangle, then the triangles are congruent.** (Tick marks and arcs indicate congruent sides and angles.)

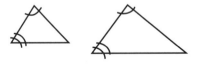

- *AA similarity triangle postulate*: **If two angles of one triangle are congruent to two angles of another triangle, then the triangles are similar.**

If two corresponding angles in each of the two triangles are equal to each other, then the third angles will be equal because the three angles in a planar triangle always sum to 180°. If all three pairs of corresponding angles in two triangles are equal to each other, the two triangles are *similar triangles*. The corresponding sides of *similar triangles* have the same proportion, but one triangle is larger than the other triangle. (Remember, congruent triangles have both corresponding angles and corresponding sides equal.)

1.5 Chapter 1 Summary and Highlights

• This chapter provides a list of terms in Section 1.1. You should be familiar with frequently used symbols including: Congruent ≅, similar ~, parallel ||, triangle Δ, perpendicular ⊥, measure of *m*, and angle *L*.

• *Deductive reasoning* is a process used to work through *proofs of theorems* and *geometric statements*. It begins with *general assumptions* and truths and derives *specific conclusions*. The process of deductive reasoning can be structured in the form of a *syllogism*, which is composed of: A *general statement* called the *major premise*; a *particular statement* called a *minor premise*; and a *conclusion* or deduction (which applies the general statement to the particular statement).

• *Inductive reasoning* begins with specific assumptions and truths and concludes something general, which may or may not always hold true in all cases. For example, we observe that most birds fly, and may conclude that *all* birds fly; however, this is not always true.

• A *theorem* is a statement that is generally proven using the process of *deductive reasoning* and building to the final proof statement using *definitions, axioms, postulates,* and other *proved theorems.* Theorems are written as conditional statements that are comprised of two parts—the *given* or *hypothesis*, which is what is known and the *conclusion* or *proof,* which is what needs to be proven. Theorems can be written in the form of either an *if-then* or a *subject-predicate* (declarative) *conditional statement.*

• A *proof* generally involves reasoning deductively using definitions, axioms, postulates, and proved theorems in order to prove a particular statement or theorem. The formal process of *proving a theorem* is:

(1) State the theorem and identify what is *given* (the *hypothesis*) and what is to be *proved* (the *conclusion*).

(2) Sketch a figure that illustrates the situation you are proving.

(3) Lay out a plan, which is not technically part of a formal proof.

(4) Create a two-column listing comprising a logical flow of statements building to the final proof statement of what is proved on the left, and reasons why each of these statements is valid on the right.

• *Theorems* are generally written and proved using deductive reasoning and already accepted statements as ***direct proofs***. In a direct proof, there are premises that lead to a conclusion. If a direct proof is not practical, an indirect proof may be used. An ***indirect proof*** begins with the *negative* of the desired conclusion and continues until reaching a contradiction between one of the statements and the negative of the desired proof. The contradiction establishes that the negative of the desired conclusion cannot be true, and the desired conclusion *must* be true.

• Selected axioms and postulates: ***Transitive axiom***: Things equal to the same or other things are equal to each other. ***One line postulate***: One and only one straight line can be drawn between any two points. ***Parallel postulate***: Through a given point not on a line, there is one and only one line parallel to that line. ***Parallel transversal postulate***: If two parallel lines are cut by a transversal, then corresponding angles are congruent. Also, alternate interior angles are congruent, vertical angles are congruent, and adjacent angles are supplementary and sum to 180°. ***SSS triangle congruence postulate***: If the three sides of one triangle are congruent to the three sides of another triangle, the triangles are congruent. ***SAS triangle congruence postulate***: If two sides and the included angle of one triangle are congruent to two sides and the included angle of another triangle, the triangles are congruent. ***ASA triangle congruence postulate***: If two angles and the included side of one triangle are congruent to two angles and the included side of another triangle, then the triangles are congruent. ***AA similarity triangle postulate***: If two angles of one triangle are congruent to two angles of another triangle, then the triangles are *similar*.

Notes

Chapter

2

Points, Lines, Planes, and Angles

Points, *lines*, and *planes* are considered *undefined* terms that are accepted as fundamental elements and used in definitions in geometry. Points, lines, and planes are described rather than defined. They are building blocks of geometry. *Angles* are formed where lines, rays, or segments meet or intersect at a point. For example, in optics, light rays form angles as they are reflected and refracted at interfaces.

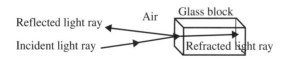

2.1 Points, Lines, and Planes

Points

• A *point* represents *position* and has no length, no width, and no height or thickness. It is represented by a dot, which is generally labeled with a capital letter. A point is often thought of as the most fundamental or basic element in geometry. All geometric figures consist of points.

Lines

• A *line* has length only, with no width, and no height or thickness. It is, in essence, a set of connected points that indefinitely extends in one dimension in both directions. A line is generated by moving a point. *A line can be straight or curved or a combination.* A line contains **collinear points**, which are points that lie in the same line. A line can be labeled by two of its points that are represented by capital letters with an arrow above them. Another label for a line is a lowercase letter. Lines are drawn with or without arrows.

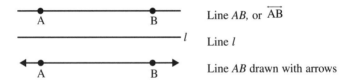

Line *AB,* or $\overrightarrow{AB}$

Line *l*

Line *AB* drawn with arrows

Postulates Pertaining to Lines

• *One line postulate*: **One and only one straight line can be drawn between any two points**.

Line *AB* is the only line that can be drawn between point A and point B.

• *Line has two points postulate*: Any line contains at least two points.

• *Two points one line postulate*: **Through any two points there is one and only one line**. (Two points determine a line.)

• *Shortest distance postulate*: **The shortest distance between any two points is the straight line that is drawn between them**.

The straight line is shorter than the broken line or curve.

• *Line intersection postulate*: Two straight lines can intersect each other at one and only one point.

Planes

• A *plane* has length and width only, with no height or thickness. A plane is a flat, two-dimensional *surface*, which is, in essence, a set of points that extend indefinitely in two dimensions. In a plane, a straight line connecting any two of its points lies completely in the plane. A plane can be represented by a capital letter preceded by the word *plane*.

Plane P

Two points and the line connecting them lie in the plane.

Postulates, Principles, and Theorems Pertaining to Planes

• *Plane has three points postulate*: Any plane contains at least three noncollinear points.

Points are called *coplanar* if and only if there is a plane that contains them.

• *Three points one plane postulate*: Through any three noncollinear points, there is one and only one plane.
 Also, through any three linear points, there is at least one plane.

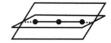

Three noncollinear points = one plane.

Three linear points = more than one plane.

• *Two points and line in plane postulate*: If two points lie in a plane, then the line joining them lies in the plane.

• *Intersection of planes postulate*: If two planes intersect, their intersection is a line.

• If a point lies outside a line, then one and only one plane contains both the line and the point.

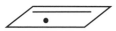

• If two lines intersect, then one and only one plane contains both lines.

• **Example:** For the cube sitting on plane P, which of the points are collinear and which are coplanar?

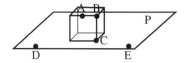

A and B are collinear and because points that lie on the same line also lie on the same plane, they are also coplanar.

B and C are collinear and because points that lie on the same line also lie on the same plane, they are also coplanar.

A and C are collinear and because points that lie on the same line also lie on the same plane, they are also coplanar.

D and E are collinear and because points that lie on the same line also lie on the same plane, they are also coplanar.

A, B, and C are coplanar, but not on plane P.

D, E, and C are coplanar on plane P.

D and C are collinear and also coplanar on plane P.

E and C are collinear and also coplanar on plane P.

Parallel Planes

• *Parallel planes* are planes that do not intersect.

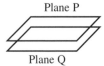

Parallel planes P and Q
never intersect.

• ***Three parallel planes theorem***: If two planes are each ***parallel*** to a third plane then the two planes are parallel to each other.

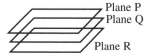

Plane P || plane R and plane Q || plane R, therefore plane P || plane Q.

• *Intersected parallel planes theorem*: If two *parallel planes* are intersected by a third plane, then the lines of intersection are parallel. (See Section 2.3 *Parallel Lines* for a proof.)

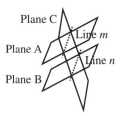

Plane A || plane B, and both planes A and B are intersected by plane C. Therefore line *m* || line *n*, where *m* and *n* are intersection lines.

Perpendicular Planes

• A *line is perpendicular to a plane* if the line is perpendicular to all of the lines that lie in the plane.

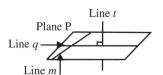

If $t \perp q$ and $t \perp m$, and q and m lie in plane P, then $t \perp$ P. (Line *t* passes through plane P, whereas *m* and *q* lie in plane P.)

• *Plane P is perpendicular to plane Q*, if plane P contains a line that is perpendicular to plane Q.

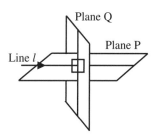

If line $l \perp$ plane Q, then plane P $\perp$ plane Q.

• *Parallel and perpendicular planes theorem*: If two planes are *perpendicular* to a third plane, then the two planes *either* are parallel to each other or intersect.

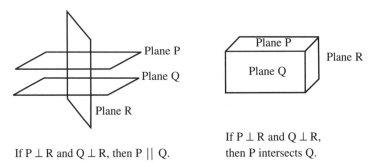

If P ⊥ R and Q ⊥ R, then P ‖ Q.

If P ⊥ R and Q ⊥ R, then P intersects Q.

• *Four points in space postulate*: Space contains at least four points not in the same plane. In other words, any four points defining space cannot line in the same plane.

2.2 Line Segments and Distance

• *A line segment is a section of a line between two points and consists of the two endpoints and all the points in between.* Line segments are generally considered to be straight (rather than curved) unless otherwise specified. A line segment is named by its endpoints. There are three line segments in the line below (left), segment *AB*, segment *BC*, and segment *AC*. A line segment is often denoted by the two letters representing its endpoints with a line over them, for example $\overline{AB}$, $\overline{AC}$, and $\overline{BC}$ or equivalently *AB*, *AC*, and *BC*. A line segment may also be labeled according to its length by a small letter such as *a* or its corresponding number.

The *length of a line segment* is the distance between its endpoints.

• Line segments on a *number line* are sections of the line that include the segment-endpoints and all points in between. A real number line is made up of *real numbers*, which are comprised of rational and irrational numbers that include natural numbers, whole numbers, integers, fractions, and decimals. All real numbers except zero are either positive or negative. All real numbers correspond to points on the *real number line*, and all points on the number line correspond to real numbers. The real number line reaches from negative infinity ($-\infty$) to positive infinity ($+\infty$), with negative numbers to the left of zero and positive numbers to the right of zero.

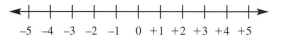

Real numbers include -0.5, -2, $5/2$, and $\pi = 3.14159...$

• When numbers are added and subtracted, we can think of moving along the number line. Begin with the first number and move to the right for positive numbers and addition, or move to the left for adding negative numbers and subtraction. (See Section 9.1 for an explanation.)

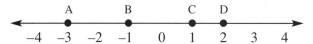

• The *ruler postulate* is used in geometry when we are working with points on a line. It allows us to measure *distance* using any unit of measure such as feet or meters. The *ruler postulate* pertains to points on the number line and states the following.

The points on a line can be numbered so that:

Every point corresponds to one and only one real number coordinate (such as zero, one, etc);

Every real number coordinate corresponds to one and only one point;

The distance between two points is the absolute value of the difference between the coordinates of the points, and every pair of points corresponds to one real number, which is the distance between the points.

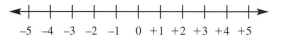

A corresponds to -3, B corresponds to -1, C corresponds to 1, and D corresponds to 2.

The **length** of segment AB is the distance between point A and point B. The distance between A and B is: $|(-1) - (-3)| = |-1 + 3| = 2$, or equivalently, $|(-3) - (-1)| = |-3 + 1| = |-2| = 2$

• *The length of a segment on a number line can be found by subtracting the coordinates of its endpoints and taking the absolute value.*

(Remember the *absolute value*, or *magnitude*, denoted by $|n|$, of a number n is defined as the distance between zero and the number on the number line and it is always positive or zero such that $|-5| = 5$.)

• Summarizing: The ***distance between two points on the number line
is the absolute value of the difference in their coordinates.*** The distance
between two points is also the *length of the line segment between them.*

• **Example:** Find length *DA* and length *CB*.

Length $DA = |2 - (-3)| = |2 + 3| = 5$
or equivalently, length $AD = |(-3) - 2| = |-5| = 5$

Length $CB = |1 - (-1)| = |1 + 1| = 2$
or equivalently, length $BC = |(-1) - 1| = |-2| = 2$

The order of subtracting doesn't matter when using absolute value.

• If a line ***segment is divided further into segments***, or there are two or
more segments together on a line, then the following are true:

1. The whole segment is equal to the sum of its parts: $AB + BC = AC$.
2. The whole segment is greater than any part: $AC > AB$ and $AC > BC$.

These principles are consistent with the following two postulates:

• ***Segment addition postulate*****: If point B is between points A and C,
then: segment** *AB* **+ segment** *BC* **= segment** *AC***, or** $\overline{AB} + \overline{BC} = \overline{AC}$.

$AB + BC = AC$

• ***Partition axiom*****:** The whole equals the sum of its parts.

If $AB = 3$, $BC = 2$, $CD = 2$
then, $AD = 3 + 2 + 2 = 7$

• **Example:** If lengths $AC = 15$, $AB = 2x$, and $BC = x$, what are lengths
AB and *BC*?

Because $AB + BC = AC$, substitute the given information and solve:

$2x + x = 15$
$x(2 + 1) = 15$
$x(3) = 15$
$x = 15/3 = 5$

Therefore, $AB = 2x = 2(5) = 10$ and $BC = x = 5$.

To check substitute into $AB + BC = AC$ or $2x + x = 15$, which results in $10 + 5 = 15$, which is correct.

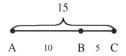

A   10   B   5   C

• *Collinear points* are points lying on the same line. Therefore if A, B, and C are collinear points having coordinates a, b, and c, then point B is between point A and point C if and only if $a < b < c$ or $a > b > c$.

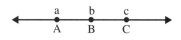

If $a < b < c$ or $a > b > c$,
then B is between A and C.

If B is between A and C,
then $a < b < c$ or $a > b > c$.

Using this, we can prove that if point B is between point A and point C, then lengths $AB + BC = AC$.

Given: Point B is between point A and point C.

Prove: $AB + BC = AC$.

Plan: Because point B is between points A and C, then $a < b < c$ or $a > b > c$. By the ruler postulate, the distance between both A and B and B and C is the absolute value of the difference in their coordinates. We combine these and substitute the lengths.

Proof:

Statements	Reasons
1. If B is between A and C, then $a < b < c$ or $a > b > c$.	1. Definition: Collinear points. (Remember, < means less than and > means greater than.)
2. If $a < b < c$, then lengths: $AB = \|a - b\| = b - a$; $BC = \|b - c\| = c - b$; $AC = \|a - c\| = c - a$.	2. Ruler postulate: The distance between two points is the absolute value of the difference between the coordinates of the points.
3. $AB + BC = (b - a) + (c - b)$ $= b - a + c - b = c - a$.	3. Substitution axiom: A quantity may be substituted for an equal quantity in any expression or equation.
4. $AB + BC = AC$.	4. Substituting from #2 and #3.

• If a ***line segment is divided into two congruent parts***, then the following is true. (Remember, *congruent* (≅) objects have the same size, shape, and measure.)

1. The two congruent parts have equal length.
   (*AB* = *BC* in the figure below.)
2. The point where the segment is divided is the ***midpoint*** of the segment.
   (Point B in the figure below.)
3. A line drawn through the midpoint ***bisects*** the segment.
   (Line *l* bisects segment *AC* at midpoint B.)

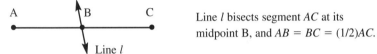

Line *l* bisects segment *AC* at its midpoint B, and *AB* = *BC* = (1/2)*AC*.

• ***Definition***: The ***midpoint of a line segment*** is halfway between each endpoint, and is therefore equidistant to the endpoints and divides the segment into two *congruent segments*. The midpoint of a line segment bisects the segment.

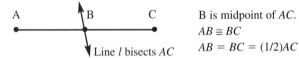

B is midpoint of *AC*.
AB ≅ *BC*
AB = *BC* = (1/2)*AC*

• ***Definition***: A ***bisector of a line segment*** is a line segment, ray, or plane that intersects the segment at its midpoint. In the above figure, line *l* bisects segment *AC* at midpoint B.

• **Example:** What is the *midpoint* of segment *AD*?

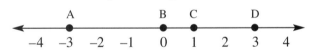

The midpoint is half of the length, therefore,
length *AD* = *DA* = 3 − (−3) = 3 + 3 = 6.

Half of the length of *AD* = (1/2)6 = 3.

Therefore, the midpoint is 3 spaces from either endpoint, A or D, so we can add 3 to −3 or subtract 3 from +3.

The result is 0, which is point B.

Note that ***the midpoint can also be obtained from taking the average of the endpoints***. (Remember, to find the average of *n* numbers, divide the *sum* of the numbers by *n*, the number of numbers.)

For this example the two endpoints are, A = −3 and D = 3:
(3 + (−3))/2 = 0/2 = 0, which is point B.

• The *midpoint postulate* states that any straight line segment has one and only one midpoint.

A   B   C   Segment *AC* has B as its only midpoint.
                   Segments *AB* and *BC* have equal lengths.

• The *midpoint theorem* states that **the midpoint of a line segment divides the segment into two sections that are each half the length of the original segment.**

Following is a proof: A   B   C

Given: Point B is the midpoint of *AC*.

Prove: *AB* = (1/2)*AC* and *BC* = (1/2)*AC*.

Plan: Given that point B is the midpoint of *AC*, use the definition of a midpoint, *AB* = *BC*, combined with the segment addition postulate, which states that if point B is between points A and C, then *AB* + *BC* = *AC*.

Proof:

Statements	Reasons
1. Point B is the midpoint of *AC*.	1. Given.
2. *AB* = *BC*.	2. The definition of a midpoint.
3. *AB* + *BC* = *AC*.	3. Segment addition postulate: If B is between A and C, then *AB* + *BC* = *AC*.
4. *AB* + *AB* = *AC*.	4. Substitute *AB* = *BC* from #2 into #3.
5. *AB* + *AB* = *AC* => 2*AB* = *AC* is equivalent to *AB* = (1/2)*AC*.	5. Rearrange #4 using algebra.
6. *BC* = (1/2)*AC*.	6. Substitute *AB* = *BC* from #2 into #5.

2.3 Parallel Lines

• *Parallel lines* **are straight lines that lie in the same plane and do not intersect even at extended lengths.** Parallel lines remain equidistant. Railroad tracks are familiar examples of parallel lines. The symbol for parallel lines is ||. If $\overrightarrow{AB}$ is parallel to $\overrightarrow{CD}$, we write $\overrightarrow{AB}$ || $\overrightarrow{CD}$.

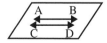

Parallel lines

• Note that *skew lines* are straight lines that do *not* lie in the same plane and do not intersect.

Skew lines

• Line segments and rays can also be parallel. In addition, a line can be parallel with a plane if they do not intersect.

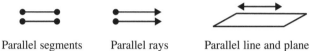

Parallel segments   Parallel rays   Parallel line and plane

• The presence and properties of parallel lines are distinguishing features of *Euclidean* vs. *non-Euclidean geometry*. In Euclidean geometry the ***parallel line postulate*** holds true, that is, *through a given point not on a line, there is ONE AND ONLY ONE line parallel to that line*. In *hyperbolic geometry* (a non-Euclidean geometry), the parallel postulate becomes: *through a given point not on a line, there is MORE THAN ONE line parallel to that line*. In *spherical geometry* (a non-Euclidean geometry) the parallel postulate becomes: *Through a given point not on a line, there ARE NO lines parallel to that line.*

• *Parallel (line) postulate*: **Through a given point not on a line, there is *one and only on*e line parallel to that line**.

Parallel line through point
Line

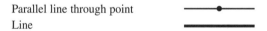

• When there are two or more lines in a plane, and they are both intersected by a third line also in the plane, this third line is called a ***transversal***. If two lines are intersected by a transversal line, there are eight angles formed. We will describe these angles for the case of two parallel lines intersected by a transversal in the following pages.

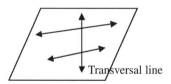

Transversal line

• A ***transversal*** is a straight line that crosses two or more parallel or non-parallel lines. The use of a transversal is helpful when we are determining angles associated with parallel lines because there are certain useful properties pertaining to the angles surrounding a transversal. These properties are described by the *parallel transversal postulate* and related principles listed in the following paragraphs. ***These principles are used throughout geometry and in proofs so pay special attention.***

- *Parallel transversal postulate*: **If two parallel lines are cut by a transversal, then corresponding angles are congruent.**

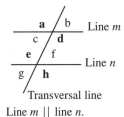

Line *m* || line *n*.

Corresponding angles: $La \cong Le$, $Lb \cong Lf$, $Lc \cong Lg$, and $Ld \cong Lh$.

Alternate interior angles: $Lc \cong Lf$ and $Ld \cong Le$.

Vertical angles: $La \cong Ld$, $Lb \cong Lc$, $Le \cong Lh$, $Lf \cong Lg$.

Corresponding angles are: $La \cong Le$, $Lb \cong Lf$, $Lc \cong Lg$, and $Ld \cong Lh$.

***Converse of parallel transversal postulate*: If two lines are cut by a transversal, and a pair of *corresponding angles* is congruent, then the two lines are parallel.**

Corresponding angles are on the same side of the transversal and on the same sides of the parallel lines.

Interior angles are between the two parallel lines: Lc, Ld, Le, and Lf.

Exterior angles are outside the two parallel lines: La, Lb, Lg, and Lh.

- The following properties are also true for parallel lines and associated with the *parallel transversal postulate*.

- ***Transversal theorem*: If two lines are parallel and intersected by a transversal, then every pair of angles formed are either congruent or supplementary.**

- ***Alternate interior angles theorem*: If two parallel lines are cut by a transversal, each pair of *alternate interior angles* is congruent.** The alternate interior angles are interior angles on opposite sides of the transversal: $Lc \cong Lf$ and $Ld \cong Le$.

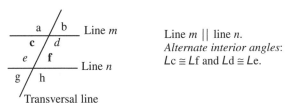

Line *m* || line *n*.
Alternate interior angles:
$Lc \cong Lf$ and $Ld \cong Le$.

***Converse*: Two lines are parallel *if*, when intersected by a transversal, a pair of *alternate interior* angles is congruent.**

• *Alternate exterior angles theorem*: If two parallel lines are cut by a transversal, each pair of *alternate exterior angles* is congruent. The alternate exterior angles are: $\angle a \cong \angle h$ and $\angle b \cong \angle g$.

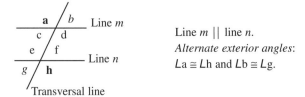

Line m || line n.
Alternate exterior angles:
$\angle a \cong \angle h$ and $\angle b \cong \angle g$.

Converse: Two lines are parallel *if*, when intersected by a transversal, a pair of *alternate exterior* angles is congruent.

• *Consecutive interior angles theorem*: **If two parallel lines are cut by a transversal, each pair of** *consecutive interior angles (on the same side of the transversal)* **is supplementary.** (Supplementary angles sum to 180°.) Supplementary consecutive interior angles are: $\angle c$ and $\angle e$, and $\angle d$ and $\angle f$.

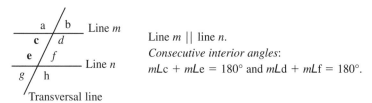

Line m || line n.
Consecutive interior angles:
$m\angle c + m\angle e = 180°$ and $m\angle d + m\angle f = 180°$.

Converse: Two lines are parallel *if*, when intersected by a transversal, a pair of *consecutive interior angles* is supplementary.

• *Consecutive exterior angles theorem*: If two parallel lines are cut by a transversal, each pair of *consecutive exterior angles (on the same side of the transversal)* is supplementary. (Supplementary angles sum to 180°.) Supplementary exterior angles are: $\angle a$ and $\angle g$, and $\angle b$ and $\angle h$.

a / b	Line m
c / d	
e / f	Line n
g / h	

Transversal line

Line m || line n.
Consecutive exterior angles:
$m\angle a + m\angle g = 180°$ and $m\angle b + m\angle h = 180°$.

Converse: Two lines are parallel *if*, when intersected by a transversal, a pair of *consecutive exterior angles* is supplementary.

• When two lines intersect, *vertical angles* are formed. ***Vertical angles*** **are the angles across the intersection point of any two intersecting lines and are congruent to each other.** In this figure, the vertical angles are: $\angle a \cong \angle d$, $\angle b \cong \angle c$, $\angle e \cong \angle h$, and $\angle f \cong \angle g$.

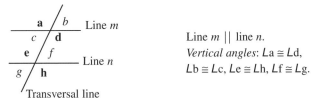

Line $m \parallel$ line n.
Vertical angles: $\angle a \cong \angle d$,
$\angle b \cong \angle c$, $\angle e \cong \angle h$, $\angle f \cong \angle g$.

• *Side note*: When two lines intersect each other, there are four angles formed that sum to 360° ($m\angle a + m\angle b + m\angle c + m\angle d = 360°$). The sum of the *adjacent (supplementary) angles*, $\angle a$ and $\angle b$, $\angle c$ and $\angle d$, $\angle a$ and $\angle c$, and $\angle b$ and $\angle d$ are each 180°. The angles opposite to each other are *vertical angles* and they are congruent: $\angle a \cong \angle d$ and $\angle b \cong \angle c$.

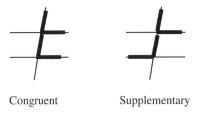

• If the sides of two angles are respectively parallel to each other, the two angles are either congruent or supplementary.

Congruent   Supplementary

• ***Parallel to same line theorem*****:** In a plane, lines are parallel if they are each parallel to the same line.

Line l Lines m and n are both
Line m parallel to line l
Line n and also to each other.

Converse**:** If lines are parallel, a line parallel to one of them is also parallel to the others.

• *Parallel to perpendicular theorem*: In a plane, lines are parallel if they are each perpendicular to the same transversal line. (□ denotes perpendicular.)

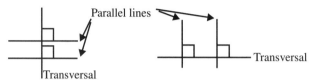

Converse: In a plane, if lines are parallel, a transversal line perpendicular to one of them is also perpendicular to the other.

• **Example:** Find La and Lb.

From the figure, we observe that La and the 60° angle are consecutive exterior angles and are therefore supplementary, which means that:
180° = 60° + mLa, therefore mLa = 180° – 60° = 120°

Next, La and Lb are also supplementary angles, therefore:
180° = mLa + mLb, or mLb = 180° – mLa = 180° – 120° = 60°
Therefore, mLa = 120° and mLb = 60°.

Another approach is to observe from the figure that Lb and the 60° angle are alternate exterior angles and are therefore congruent, which indicates that: mLb = 60°

Because La and Lb are supplementary angles, then:
180° = mLa + mLb, or mLa = 180° – mLb = 180° – 60° = 120°
Again, we determine: mLa = 120° and mLb = 60°.

There is usually more than one approach to solving these types of problems.

• **Example:** Prove the *intersected parallel planes theorem*, which states that if two parallel planes are intersected by a third plane, then the lines of intersection are parallel.

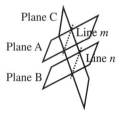

Plane P || plane Q and both P and Q are intersected by plane R. Therefore line *m* || line *n*, where *m* and *n* are intersection lines.

Following is a proof:

Given: Plane P ‖ plane Q;
        line m lies in the intersection of plane P and plane R;
        line n lies in the intersection of plane Q and plane R.

Prove: Line m ‖ line n.

Plan: Show m ‖ n because m and n don't intersect and lie in the same plane R. Lines m and n don't intersect because they lie in parallel planes.

Proof:

Statements	Reasons
1. Lines m and n each lie in plane R.	1. Given: At intersections of R with P & Q.
2. Lines m and n are coplanar.	2. Def: Coplanar lines lie in the same plane.
3. Plane P ‖ plane Q.	3. Given.
4. Line m lies in plane P; line n lies in plane Q.	4. Given.
5. Planes P and Q don't intersect.	5. Def: Parallel planes don't intersect.
6. Line m and line n don't intersect.	6. m and n lie in planes that don't intersect.
7. Line m ‖ line n.	7. Def. of parallel lines: Lines m and n don't intersect and each lie in the same plane R.

• **Example:** Given that ∠a is supplementary to ∠g, prove that line m and line n are parallel.

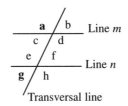

Line m

Line n

Transversal line

Given: ∠a is supplementary to ∠g.

Prove: Line m ‖ line n.

Plan: Because ∠a is supplementary to ∠g, and ∠e is supplementary to ∠g, then ∠a and ∠e are both supplementary to ∠g and are therefore congruent to each other. Because ∠a ≅ ∠e and they are corresponding angles, then line m ‖ line n, by the parallel transversal postulate.

Proof:

Statements	Reasons
1. La is supplementary to Lg.	1. Given.
2. Le is supplementary to Lg.	2. Def: Two angles whose measures total 180° (straight line) are supplementary.
3. La ≅ Le.	3. Ls supplementary to the same L (g) are ≅.
4. Line m ‖ line n.	4. Parallel transversal postulate: Two lines are parallel if, when cut by a transversal, a pair of corresponding angles are ≅. (#3)

2.4 Perpendicular Lines

• *Perpendicular lines* are defined as follows: Two lines are perpendicular if and only if they form a right angle. We can also say that two lines, rays, or segments are *perpendicular* if they lie in the same plane and intersect at right (90°) angles.

• *Four right angles theorem*: If two lines cross each other and are *perpendicular* to each other, then four *right angles* are formed.

• *Definition*: **Perpendicular lines form right angles.**

• *Definition*: **Right angles measure 90°.**

Right	90°
Congruent	angles

• *Right angles theorem*: **Any two right angles are congruent.**

• In the following drawing, angles a, b, c, and d each measure 90°, and the sum of the angles is: $m L$a + $m L$b + $m L$c + $m L$d = 360°. The *symbol for perpendicular* is ⊥, such that, line r ⊥ line t. Also, a small square in the inside vertex of an angle denotes that it is a right angle.

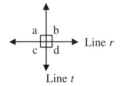

Two perpendicular lines form right angles when they cross.

Line r ⊥ line t.
La, Lb, Lc, and Ld are right angles.
La ≅ Lb ≅ Lc ≅ Ld.

• *Congruent adjacent angles theorem*: If two lines are perpendicular, then they form *congruent adjacent angles*.

Converse theorem: If two lines form *congruent adjacent angles*, then they are perpendicular.

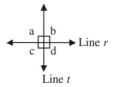

Prove the *congruent adjacent angles theorem* using the preceding figure.

Given: Line $r \perp$ line t.

Prove: $\angle$a, $\angle$b, $\angle$c, and $\angle$d are congruent angles.

Proof:

Statements	Reasons
1. Line $r \perp$ line t.	1. Given.
2. $\angle$a, $\angle$b, $\angle$c, $\angle$d are 90° angles.	2. Def: $\perp$ lines form right, 90°, angles.
3. $\angle$a, $\angle$b, $\angle$c, $\angle$d are congruent angles.	3. Def: Right angles are congruent.

Postulates Pertaining to Perpendicular Lines

• ***Perpendicular through line postulate***: One and only one perpendicular line can be drawn to or through any point *on* a line (in a plane).

Point B *on* segment *AC* has one perpendicular line, *BD*, that is drawn to it.

• ***Perpendicular point to line postulate***: One and only one perpendicular line can be drawn from or through any point *not on* a line to that line.

One line from *external* Point D is drawn perpendicular to line *AC*.

2.5 Distances and Bisectors

• The ***distance between any two geometric objects*** is always a line segment that is the shortest path or line between the two objects. The objects may be, for example, two points, a point and a line, two parallel lines, a point and a circle, or two concentric circles. The distance between two geometric objects is described in the following paragraphs.

• The *distance between two points* (X and Y) is measured by the shortest line segment between them (*XY*).

• *Shortest distance postulate*: The shortest distance between any two points is the straight line that is drawn between them.

The straight line is shorter than the bent line or the curve.

• The *distance between a point* (X) *and a line* (*AB*) is measured by the shortest line segment between them (*XY*), which is the *perpendicular from the point to the line*.

XY ⊥ *AB* and *XY* is therefore the shortest distance from X to *AB*. *XC* is *not* ⊥ to *AB* and is *not* the shortest distance from X to *AB*.

Note: □ denotes that *XY* is perpendicular to (⊥) *AB*. (See proof on page 49.)

• The *distance between two parallel lines* (*AB* and *CD*) is measured by *a perpendicular line segment between them* (*XY*).

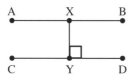

AB || *CD*; *AB* ⊥ *XY*; *CD* ⊥ *XY*. Distance between *AB* and *CD* is length *XY*. Every perpendicular segment joining *AB* and *CD* has the same length.

• *Parallel and perpendiculars theorem*: If two lines are parallel, every perpendicular segment joining them has the same length. (See proof on page 50.)

• The *distance between a point* (X) *and a circle* is measured by the shortest line segment between them (*XY*), which is on the segment *CX* (an extension of a radius) between the point and the circle.

The shortest distance between point X and the circle is *XY*, which is on the extension of a radius line.

• The *distance between two concentric circles* (each with center C) is measured by the shortest line segment between them (*XY*), which is on the larger radius between the two circles.

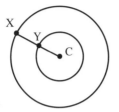

The shortest distance between two concentric circles is the section of the radius of the larger circle that lies between the two circles.

Principles related to distance: A *perpendicular bisector of a segment*

• *Perpendicular bisector/endpoints theorem*: If a point lies on the *perpendicular bisector* of a line segment, then the point is *equidistant* from the endpoints of a line segment.

Converse of perpendicular bisector/endpoints theorem: If a point is equidistant from the endpoints of a line segment, then the point lies on the perpendicular bisector.

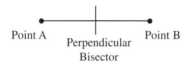

Note: A *locus* is the set of points that satisfy a certain condition or conditions. In this case, a perpendicular bisector represents the locus of points equidistant between two points on the line connecting them. The locus of points of the perpendicular bisector satisfy the conditions that require all points to be equidistant between two given points.

• A line segment and its perpendicular bisector may be part of any geometric figure, such as a rectangle or triangle.

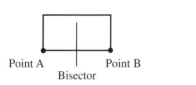

In $\triangle ABC$, $AC \cong BC$.

• The line (*XY*) joining the vertices of two isosceles triangles (△AXB and △AYB) having a common base (*AB*) is the *perpendicular bisector* of the base.

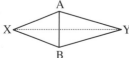

△AXB and △AYB are isosceles.
XY is the perpendicular bisector of *AB*.
$XA \cong XB$ and $YA \cong YB$.

Principles related to distance: A *bisector of an angle*

• *Point on angle bisector theorem*: If a point is on the *bisector of an angle*, then it is *equidistant* from the sides of the angle.

Converse of point on angle bisector theorem: If a point is equidistant from the sides of an angle, then it is on the bisector of the angle.

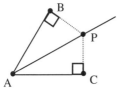

If *AP* is the bisector of ∠BAC, then *BP* ≅ *PC*.

If *BP* ≅ *PC*, then *AP* is the bisector of ∠BAC.

If ∠B and ∠C are right angles, then *BP* and *CP* represent the **distances** between P and the sides of the angle. (Remember, the distance between a point and a line is measured by the shortest line segment between them, which is the perpendicular segment drawn from the point to the line.)

• *Side note*: The **locus** of points satisfying the conditions that they be *equidistant* from the sides of a given *angle,* is the *bisector* of that angle.

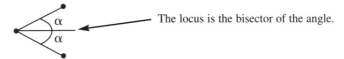

The locus is the bisector of the angle.

Principles related to distance: *Perpendicular bisectors of the sides of a triangle*

• *Perpendicular bisectors concurrent theorem*: The perpendicular bisectors of the sides of a triangle are concurrent (meet at point) and intersect at a point that is equidistant from the three vertices of the triangle.

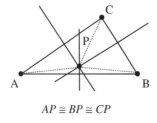
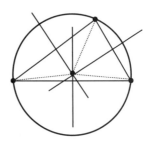

AP ≅ *BP* ≅ *CP*

The perpendicular bisectors of the sides of a *triangle* intersect at the center (*circumcenter*) of a *circle that circumscribes the triangle* and has the distance from center to each vertex as its radius (dashed).

Principles related to distance: *Bisectors of the angles of a triangle*

• *Equidistant angle bisectors theorem*: The bisectors of the angles of a triangle are concurrent (meet at point) and intersect at a point that is equidistant from the three sides of the triangle.

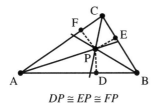 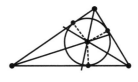

$DP \cong EP \cong FP$

The bisectors of the angles of a triangle intersect at the center (*incenter*) of a *circle that can be inscribed in the triangle*. The inscribed circle has a radius (dashed lines) equal to the length of a line drawn perpendicular to any side from its center.

• **Example:** Prove that the perpendicular segment (*XY*) from a point (X) to a line (*AB*) is shorter than any non-perpendicular segment (*XC*) between the same point (X) and same line (*AB*).

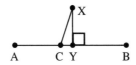

☐ denotes right angles are formed because *XY* is perpendicular (⊥) to *AB*.

Given: Line *AB* with external point X;
segment *XY* is perpendicular to line *AB*;
segment *XC* is a non-perpendicular from point X to line *AB*.

Prove: Segment *XY* is shorter than segment *XC*.

Proof:

Statements	Reasons
1. *AB* with external pt. X; *XY* ⊥ *AB*.	1. Given.
2. ∠CYX is a right angle.	2. Definition: ⊥ lines form right angles.
3. ∠XCY is an acute angle.	3. Non-right angles of a right triangle (ΔXYC) are acute and always < 90°.
4. ∠XCY < ∠CYX.	4. Def: Acute angles are less than < 90°.
5. *XY* is shorter than *XC*.	5. In a triangle (ΔXYC) the side opposite the smaller ∠ is shorter than side opposite the larger angle. (∠XCY < ∠CYX, so *XY* < *XC*)

The distance between a point and a line is measured by the shortest (perpendicular) line segment between them.

• **Example:** Prove the theorem which states that if two lines are parallel, every perpendicular segment joining them has the same length.

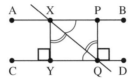

Given: Line $AB \parallel$ line CD; $XY \perp CD$; $PQ \perp CD$.

Prove: Segment $XY \cong$ segment PQ.

Proof:

Statements	Reasons
1. $XY \perp CD$; $PQ \perp CD$; $AB \parallel CD$.	1. Given.
2. Draw XQ transversal.	2. Two points, X and Q, determine a line.
3. $\angle PXQ \cong \angle XQY$; $\angle YXQ \cong \angle PQX$.	3. If two $\parallel$ lines are cut by a transversal, then alternate interior $\angle$s are $\cong$.
4. $XY \parallel PQ$.	4. If two lines, XY & PQ, are $\perp$ to a third line, CD, then they are $\parallel$ to each other.
5. Segment $XQ =$ segment XQ.	5. Identity/reflexive axiom: Self equals self.
6. $\triangle XQY \cong \triangle QXP$.	6. ASA congruence post: If two $\angle$s and included side of one $\triangle$ are $\cong$ to corresponding parts of other $\triangle$, the $\triangle$s are $\cong$. (#3&#5)
7. Segment $XY =$ segment PQ.	7. Corresponding sides of $\cong \triangle$s are $\cong$.

2.6 Rays and Angles

• A *ray* has an endpoint at one end and extends indefinitely in the other direction. Rays are *named* by the letter representing the *endpoint* and any other point on the ray. The following two rays can both be called ray AB, and they each consist of one endpoint A and the set of all points in the direction of the arrow. The symbol for a ray is an arrow above the two letters representing the ray. The two rays below can be labeled $\overrightarrow{AB}$ or $\overrightarrow{BA}$ and $\overrightarrow{BA}$ or $\overrightarrow{CA}$ or $\overrightarrow{AB}$ or $\overrightarrow{AC}$.

• If two straight rays or segments meet at their endpoints, an *angle* is formed. If two straight lines cross each other at a point, four *angles* are formed. The point where the endpoints meet or lines cross is called the *vertex* of the angle, and the sides of the angle are called the *sides*, or sometimes the *rays*, of the angle.

Ray AB and ray AC are joined at point   Intersecting lines
A to form angle BAC with vertex A.   form four angles.

• The *symbol for an angle* is L. The angle above can be called LBAC (where the middle letter corresponds to the vertex) or simply LA (the vertex point). Angles are labeled by letters, numbers, or Greek letters such as α, β, δ, ϕ, and θ inside the angle.

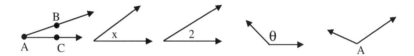

The above angles are labeled LBAC (or LCAB, or LA), Lx, L2, $L\theta$, and LA. Many angles can be labeled with more than one choice. For example, the first angle can be labeled LBAC, LCAB, or LA. When labeling an angle choose a name that will most easily allow someone else to identify it.

• An *angle* can be formed by rotating a ray around its endpoint. If an angle is formed by rotating a ray counterclockwise, a *positive angle* results. If an angle is formed by rotating a ray clockwise, a *negative angle* results.

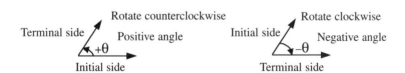

• The *measure of an angle* is the separation between the sides in degrees or radians and is often denoted by the letter m preceding the angle symbol: mLBAC.

mLBAC in degrees or radians

• Angles are measured in **degrees** or **radians**. The symbol for degrees is °, and radians is often shortened to *rad*. Radians are called circular measures, and are often used in the context of angles rotated about the origin of a coordinate system and in trigonometric functions. A full circle measures 360° or 2π radians. Radians are commonly used in trigonometry and degrees are generally used in introductory geometry. *Degrees* can be divided into **minutes** (denoted by ') and **seconds** (denoted by ").

$1° = 60$ minutes: There are 60 *one-minute size* angles in a 1° angle.

1 minute $= 60$ seconds: There are 60 *one-second size* angles in a one-minute angle.

1 minute $= 1/60$th of a degree

1 second $= 1/60$th of a minute $= 1/3600$th of a degree

Example: Convert from degrees/minutes/seconds to degrees.

$30°15'22'' = 30° + 15/60 + 22/3600 = 30° + 0.25 + 0.006 = 30.256°$

Check: $30.256° = 30° + 0.256(60') = 30° + 15.36'$
$= 30° + 15' + 0.36(60'') = 30° + 15' + 22'' = 30°15'22''$

Example: Convert from degrees to degrees/minutes/seconds.

$63.23° = 63° + 0.23(60') = 63°13.8' = 63° + 13' + 0.8(60'') = 63°13'48''$

Check: $63°13'48'' = 63° + 13/60 + 48/3600 = 63° + 0.217 + 0.0133$
$= 63.23°$

• A **protractor** is used to find the *measure of an angle in degrees*. Simple protractors can be half-circles with a degree range from 0° to 180° or they can be complete circles with a degree range from 0° to 360°. Protractors can also have two numbered scales increasing in opposite directions so that they can be used to measure an angle beginning with either side. The location of *rays* on a protractor is referred to as its coordinate.

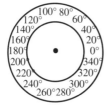

Full rotation circular protractor

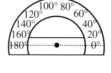

Half rotation semicircle protractor

A protractor is most easily used by placing it on the angle so that the *center point of the protractor coincides with the vertex of the angle* and the *0° line of the protractor coincides with one side of the angle*. The *measure of the angle* is determined by moving along the scale beginning at 0° and reading the degree measure of the angle where the non-zero side of the angle intersects the scale on the protractor. In the following drawing, there are two angles being measured. If the right side of each angle is along 0°, then the measures of the angles are 30° and 120°.

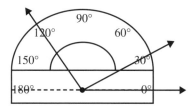

The length of the sides of an angle do not affect the measure of an angle—only the number of degrees between the two sides.

• It is also possible to measure angles using a protractor without placing one side on 0°. The *protractor postulate* (which is similar to the ruler postulate for line segments) provides the basis for any angle measurement using a protractor and states the following.

Protractor postulate: The *rays* can be numbered (for a half-rotation protractor) so that:

Every real number beginning with 0° and ending with 180° corresponds to one and only one ray; every real number coordinate corresponds to one and only one point; every pair of rays corresponds to one real number, which is the measure of the angle they determine; and the measure of the angle is the *absolute value* of the difference between the coordinates of its rays. (Remember, the *absolute value,* denoted by $|n|$ of a number n, is always positive or zero, never negative, such that $|-5| = 5$.)

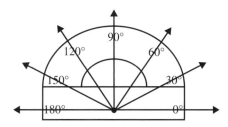

• **Example:** Estimate $mLAOB$, $mLAOC$, and $mLBOC$.

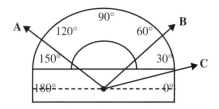

Using the protractor postulate, especially the phrase, *the measure of the angle is the absolute value of the difference between the coordinates of its rays*, we can *estimate* the measures of each angle. First, estimate the values of A, B, and C as A = 140°, B = 45°, and C = 15°.

$$m\angle AOB = |140° - 45°| = |95°| = 95°$$

$$m\angle AOC = |140° - 15°| = |125°| = 125°$$

$$m\angle BOC = |45° - 15°| = |30°| = 30°$$

• Angles can be added using the **angle addition postulate**, which states that **if point B (on ray VB) is in between point A and point C on the interior of angle AVC, then: $m\angle AVB + m\angle BVC = m\angle AVC$**

$$m\angle AVB + m\angle BVC = m\angle AVC$$

• **Example:** If $m\angle AVB = 35°$ and $m\angle BVC = 20°$, what is $m\angle AVC$?

Using the angle addition postulate:

$$m\angle AVC = m\angle AVB + m\angle BVC = 35° + 20° = 55°$$

• An *angle bisector* is a line, ray, or segment that divides an angle into two congruent angles.

• *Definition*: The two angles formed by a bisector are congruent.

• *Angle bisector postulate*: Any angle has one and only one bisector.

The bisector for $\angle ABC$ is ray *BX*.
$\angle a \cong \angle a$, or equivalently, $m\angle a = m\angle a$.

• *Angle bisector theorem*: **A ray that bisects an angle divides it into two angles half the measure of the angle.** If ray *BX* is the bisector of $\angle ABC$, then: $m\angle ABX = (1/2)m\angle ABC$ **and** $m\angle XBC = (1/2)m\angle ABC$

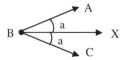

Proof of *angle bisector theorem*:

Given: Ray BX is the bisector of LABC.

Prove: mLABX $= (1/2)mL$ABC and mLXBC $= (1/2)mL$ABC.

Proof:

Statements	Reasons
1. Ray *BX* is the bisector of LABC.	1. Given.
2. LABX $\cong L$XBC.	2. Def: Two Ls formed by bisector are $\cong$.
3. mLABX $+ mL$XBC $= mL$ABC.	3. Angle addn post: If X is betw. A and C on LABC, then mLABX $+ mL$XBC $= mL$ABC.
4. mLABX $+ mL$ABX $= mL$ABC.	4. Substitution from #2 and #3.
5. $2mL$ABX $= mL$ABC.	5. Rearrange #4 using algebra.
6. mLABX $= (1/2)mL$ABC.	6. Rearrange #5 using algebra.
7. mLXBC $= (1/2)mL$ABC.	7. Substitution from #2 & #6.

• ***One bisector theorem*:** An angle that is not a straight angle has one and only one bisector.

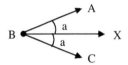

Ray *BX* is the only bisector of LABC.

Types of Angles

• A ***180° angle***, also called a ***straight angle***, has its sides lying in a straight line.

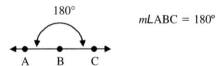

mLABC $= 180°$

• ***Straight angle theorem*:** All straight angles are equal, or congruent.

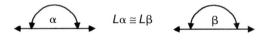

$L\alpha \cong L\beta$

• If LAVC is a *straight angle*, then mLAVB $+ mL$BVC $= 180°$.

mLAVC $= 180°$

• A *right angle* measures 90° and is often identified using a square at the interior of the vertex.

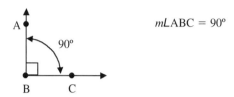

mLABC = 90°

• *Right angles (congruent) theorem*: **All right angles are congruent.**

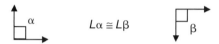

$L\alpha \cong L\beta$

Prove *right angle congruent theorem*:

Given: $L\alpha$ and $L\beta$ are right angles.

Prove: $L\alpha \cong L\beta$.

Proof:

Statements	Reasons
1. $L\alpha$ and $L\beta$ are right angles. 2. $mL\alpha$ = 90° and $mL\beta$ = 90°. 3. $L\alpha \cong L\beta$.	1. Given. 2. Definition: Right angles equal 90°. 3. Transitive axiom: Things equal to the same are equal to each other.

• **Angles smaller than 90° are called *acute angles*.**

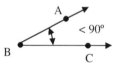

LABC is an acute angle.
Any angle measuring less than 90° is an acute angle.

• **Angles larger than 90° but less than 180° are called *obtuse angles*.**

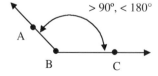

LABC is an obtuse angle.
Any angle measuring between 90° and 180° is an obtuse angle.

- Angles larger than 180° but less than 360° (a circle) are called ***reflex angles***.

> 180°, < 360°

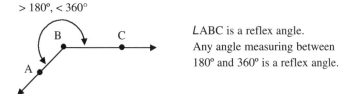

∠ABC is a reflex angle.
Any angle measuring between
180° and 360° is a reflex angle.

- If two angles have the same initial and terminal sides, they are called ***coterminal angles***. In the drawing, there are two positive angles and one negative angle that are coterminal.

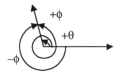

Coterminal angles can be formed by beginning at the same initial side and circling in a positive or negative direction and ending with the same terminal side. Coterminal angles can therefore contain multiples of 360° if the angle circles more than one time.

Pairs of Angles

- ***Congruent angles* have the same measure, or number of degrees.**

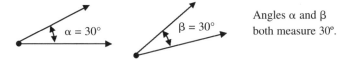

Angles α and β
both measure 30°.

∠α ≅ ∠β is also written $m\angle\alpha = m\angle\beta$, where *m* denotes *measure of*.

- If two angles lie in a plane, and have the same vertex point and share a common side, they are called ***adjacent angles***.

∠a and ∠b are adjacent angles.

• Adjacent angles can be added using the ***angle addition postulate***, which states that if point B (on ray VB) is in between point A and point C on the interior of angle AVC, then: $m\angle AVB + m\angle BVC = m\angle AVC$

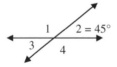

$$m\angle AVB + m\angle BVC = m\angle AVC$$

If $m\alpha = 30°$ and $m\beta = 25°$, then $m\angle AVC = 55°$.

• ***Vertical angles*** are the angles that have the same vertex, but are across the intersection point of any two intersecting lines. Vertical angles are equal in measure to each other, or congruent (see proof below). When two lines intersect, four angles are formed with the *vertical angles* as either pair of non-adjacent angles.

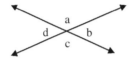

$\angle a \cong \angle c$ are vertical angles.
$\angle d \cong \angle b$ are vertical angles.

If $m\angle a = 130°$, then $m\angle c = 130°$, and if $m\angle d = 50°$, then $m\angle b = 50°$.

Note that intersecting lines produce not only the two pairs of congruent vertical angles, but also four pairs of ***adjacent supplementary angles***. The adjacent supplementary angles are: $\angle a$ and $\angle b$, $\angle b$ and $\angle c$, $\angle c$ and $\angle d$, and $\angle d$ and $\angle a$.

• **Example:** Find all four angles given $m\angle 2 = 45°$.

$\angle 2$ and $\angle 3$ are vertical angles, therefore: $m\angle 2 = m\angle 3 = 45°$
$\angle 1$ and $\angle 2$ are supplementary angles, therefore: $m\angle 1 + m\angle 2 = 180°$, or $180° - m\angle 2 = m\angle 1 = 180° - 45° = 135°$
$\angle 1$ and $\angle 4$ are vertical angles, therefore: $m\angle 1 = m\angle 4 = 135°$
Therefore, $m\angle 1 = 135°$, $m\angle 2 = 45°$, $m\angle 3 = 45°$, and $m\angle 4 = 135°$.

• ***Vertical angles theorem*****: Vertical angles are congruent.**

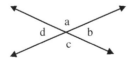

Prove *vertical angle theorem*:

Given: Two straight lines intersect forming vertical angles $\angle a$ with $\angle c$ and $\angle d$ with $\angle b$.

Prove: $\angle d \cong \angle b$.

Proof:

Statements	Reasons
1. $\angle a$ & $\angle b$ and $\angle a$ & $\angle d$ are supplementary angles.	1. Given. (Two straight lines.)
2. $m\angle a + m\angle d = 180°$ and $m\angle a + m\angle b = 180°$.	2. Supplementary angles sum to 180°.
3. $m\angle a + m\angle d = m\angle a + m\angle b$, or $m\angle a - m\angle a = m\angle d - m\angle b$.	3. Transitive: Things equal to same are equal to each other. (#2)
4. $m\angle a = m\angle a$.	4. Identity/reflexive axiom: self=self. (#3)
5. $m\angle d = m\angle b$, or $\angle d \cong \angle b$.	5. Subtraction: If equals are subtracted from equals the differences are equal. (#3, #4)

• **If the sum of any two angles equals 180°, the two angles are called *supplementary angles*.** Supplementary angles can be *adjacent and supplementary*, or they may be *non-adjacent supplementary* angles. *Adjacent supplementary angles* have a common side and vertex and sum to 180°. *Non-adjacent supplementary angles* also sum to 180°, but do *not* have a common side and vertex and are separate. Either of two supplementary angles is called the *supplement* of the other angle. In each drawing $m\angle a + m\angle b = 180°$.

Adjacent supplementary angles   Non-adjacent supplementary angles
$m\angle a + m\angle b = 180°$ $m\angle a + m\angle b = 180°$

Note that if one angle (of a pair of supplementary angles) is known, then the measure of its supplement can be found by subtracting the known measure from 180°. In the figure, if $\angle a$ is known, then $m\angle b = 180° - m\angle a$.

• *Definition*: **Supplementary angles sum to 180°.**

• **Example:** If $m\angle ABD = 120°$ and $m\angle ABC = 180°$, find $m\angle DBC$.

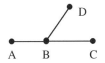

Because $\angle ABD$ and $\angle DBC$ are supplementary angles and sum to $180°$, then $m\angle ABD + m\angle DBC = m\angle ABC = 180°$, or
$m\angle ABC - m\angle ABD = m\angle DBC = 180° - 120° = 60°$
Therefore, $m\angle DBC = 60°$.

• *Adjacent supplementary angle theorem*: If two adjacent angles have their non-common sides lying in the same straight line, then the adjacent angles are supplementary.

It can also be said that adjacent angles are supplementary if their non-adjacent sides lie in the same straight line.

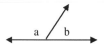

$\angle a$ and $\angle b$ are adjacent and supplementary and sum to $180°$.

• *Definition*: Two angles that have a common side and their other sides form a line (which are opposite rays) are called a *linear pair*.

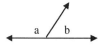

$\angle a$ and $\angle b$ are a linear pair and are supplementary and sum to $180°$.

• *Linear pair theorem*: If two angles are a *linear pair*, then they are supplementary.

• *Linear pair/right angle theorem*: If two angles in a linear pair are congruent, then each is a right angle.

Similarly, supplementary angles are congruent if they are each right angles.

$\angle ABD$ is supplementary to $\angle DBC$,
$\angle ABD$ and $\angle DBC$ form a linear pair,
$\angle ABD \cong \angle DBC$, therefore,
$m\angle ABD = 90°$ and $m\angle DBC = 90°$.

(Remember, a linear pair is two angles that have a common side and their other sides form a straight line. Also, if two angles form a linear pair, they are supplementary.)

• *Supplementary to same angle theorem*: **If two angles are supplementary to the same angle, (or to congruent angles), then they are congruent to each other**.
(See proof following.)

This theorem is a combination of the following principles:

1. Supplements of the same angle are congruent to each other.

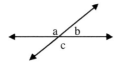

If $\angle$a is a supplement of $\angle$b
and $\angle$c is a supplement of $\angle$b,
then $\angle$a and $\angle$c are congruent.

2. Supplements of congruent angles are congruent to each other.

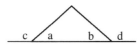

If $\angle$a is a supplement of $\angle$c,
$\angle$b is a supplement of $\angle$d,
and $\angle$a and $\angle$b are congruent,
then $\angle$c and $\angle$d are congruent.

• Proof of *supplementary to same angle theorem*:

Given: $\angle$a and $\angle$b are supplementary; $\angle$c and $\angle$d are supplementary; $\angle$a $\cong$ $\angle$c.

Prove: $\angle$b $\cong$ $\angle$d.

Proof:

Statements	Reasons
1. $\angle$a and $\angle$b are supplementary; $\angle$c and $\angle$d are supplementary.	1. Given.
2. $m\angle a + m\angle b = 180°$; $m\angle c + m\angle d = 180°$.	2. Definition: Supplementary angles sum to 180°.
3. $m\angle a + m\angle b = m\angle c + m\angle d$.	3. Transitive: Things equal to the same are equal to each other. (#2)
4. $m\angle a = m\angle c$, or $\angle$a $\cong$ $\angle$c.	4. Given.
5. $m\angle a + m\angle b = m\angle a + m\angle d$.	5. Substitute #3 with #4.
6. $m\angle b = m\angle d$, or $\angle$b $\cong$ $\angle$d.	6. Subtraction: If equals are subtracted from equals the differences are equal. (#5)

• **If the sum of any two angles is 90°, the two angles are called *com-plementary angles*.** Complementary angles can be *adjacent and complementary,* or they may be *non-adjacent complementary* angles. *Adjacent complementary angles* have a common side and vertex and sum to 90°. *Non-adjacent complementary angles* also sum to 90°, but do not have a common side and vertex and are separate. Either of two complementary angles is called the ***complement*** of the other angle. In each drawing, $m\angle a + m\angle b = 90°$.

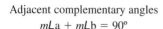

Adjacent complementary angles   Non-adjacent complementary angles
$m\angle a + m\angle b = 90°$ $m\angle a + m\angle b = 90°$

For complementary angles, if one angle is known the measure of its complement can be found by subtracting the known measure from 90°. If $\angle a$ is known, then $m\angle b = 90° - m\angle a$.

• *Definition*: **Complementary angles sum to 90°.**

• **Example:** If $m\angle ABD = 50°$ and $m\angle DBC = 40°$, find $\angle ABC$.

From the figure, we can see that $\angle ABD$ and $\angle DBC$ are adjacent and using the *angle addition postulate*: $m\angle ABD + m\angle DBC = m\angle ABC$

Substitute the measurements given: $50° + 40° = 90°$

Therefore, $\angle ABC$ is a right angle (measuring 90°), and angles $\angle ABD$ and $\angle DBC$ are adjacent and complementary angles.

• *Complementary to same angle theorem*: **If two angles are complementary to the same angle, (or to congruent angles), then they are congruent to each other.** In other words, complements of the same (or congruent) angles are congruent.

This theorem is a combination of the following principles:

1. Complements of the same angle are congruent to each other.

If $\angle a$ is a complement of $\angle b$ and $\angle c$ is a complement of $\angle b$, then $\angle a$ and $\angle c$ are congruent.

2. Complements ($\angle c$ and $\angle d$) of congruent angles ($\angle a$ and $\angle b$, respectively) are congruent to each other.

If $\angle a$ is a complement of $\angle c$, $\angle b$ is a complement of $\angle d$, and $\angle a$ and $\angle b$ are congruent, then $\angle c$ and $\angle d$ are congruent.

• Proof of *complementary to same angle theorem*:

Given: $\angle$a and $\angle$b are complementary; $\angle$c and $\angle$d are complementary; $\angle$a $\cong \angle$c.

Prove: $\angle$b $\cong \angle$d.

Proof:

Statements	Reasons
1. $\angle$a and $\angle$b are complementary; $\angle$c and $\angle$d are complementary.	1. Given.
2. $m\angle$a + $m\angle$b = 90°; $m\angle$c + $m\angle$d = 90°.	2. Def: Complementary $\angle$s sum to 90°.
3. $m\angle$a + $m\angle$b = $m\angle$c + $m\angle$d.	3. Transitive: Things = same = each other.
4. $m\angle$a = $m\angle$c, or $\angle$a $\cong \angle$c.	4. Given.
5. $m\angle$a + $m\angle$b = $m\angle$a + $m\angle$d.	5. Substitute #3 with #4.
6. $m\angle$b = $m\angle$d, or $\angle$b $\cong \angle$d.	6. Subtraction of #5.

• ***Perpendicular/complementary theorem***: If the non-adjacent sides of two acute adjacent angles are *perpendicular* to each other, then the angles are *complementary*.

$\angle$ABD is adjacent to $\angle$DBC and $AB \perp BC$ therefore, $\angle$ABD and $\angle$DBC are complementary.

Proof of theorem:

Given: $AB \perp BC$.

Prove: $\angle$ABD and $\angle$DBC are complementary.

Proof:

Statements	Reasons
1. $AB \perp BC$.	1. Given.
2. $m\angle$ABC = 90°.	2. Definition: $\perp$ lines form right, 90° angles.
3. $m\angle$ABD + $m\angle$DBC = $m\angle$ABC.	3. Angle addition postulate.
4. $m\angle$ABD + $m\angle$DBC = 90°.	4. Substitute from #2 and #3.
5. $\angle$ABD & $\angle$DBC are complementary.	5. Def: Complementary angles sum to 90°.

• *Intersecting lines*: If two *lines intersect* each other at a point, there are four angles formed. In the following diagram, when line *m* and line *n* intersect, four angles are formed (∠a, ∠b, ∠c, and ∠d), and the measures of the angles sum to 360°. The sum of the *adjacent angles* (∠a and ∠b, ∠c and ∠d, ∠a and ∠c, and ∠b and ∠d) are each 180°. The angles opposite to each other (∠a and ∠d and angles ∠b and ∠c) are *vertical angles*, and they are congruent ($m∠a = m∠d$ and $m∠b = m∠c$).

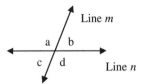

2.7 Chapter 2 Summary and Highlights

• A *point* represents position and has no length, no width, and no height or thickness. A *line* has length only, with no width, and no height or thickness. A *plane* has length and width only, with no height or thickness.

• *One line postulate*: One and only one straight line can be drawn between any two points. *Two points one line postulate*: Through any two points, there is one and only one line. *Shortest distance postulate*: The shortest distance between any two points is the straight line that is drawn between them.

• The *distance between two points on the number line is the absolute value of the difference in their coordinates*. The distance between two points is also the *length of the line segment between them*.

• *Segment addition postulate*: If point B is between points A and C, then: Segment *AB* + segment *BC* = segment *AC*

• The *midpoint theorem*: The midpoint of a line segment divides the segment into two sections that are each half the length of the original segment.

• *Parallel lines* are straight lines that lie in the same plane and do not intersect, even at extended lengths. Parallel lines remain equidistant.

• *Parallel postulate*: Through a given point not on a line, there is *one and only one* line parallel to that line.

If two lines are parallel and intersected by a *transversal*, then every pair of angles formed are either congruent or supplementary.

• *Parallel transversal postulate*: If two parallel lines are cut by a transversal, then *corresponding angles are congruent*, and in addition: *each pair of alternate interior angles is congruent*, and *each pair of consecutive interior angles is supplementary*.

• If two straight rays or segments meet at their endpoints, an ***angle*** is formed. If two straight lines cross each other at a point, four *angles* are formed.

• ***Vertical angles*** are the angles across the intersection point of any two intersecting lines and are congruent to each other. *Vertical angles theorem*: Vertical angles are congruent.

• The length of the sides of an angle do not affect the measure of an angle—only the number of degrees between the two sides.

• ***Congruent angles*** have the same measure, or number of degrees.

• ***Perpendicular lines*** form right angles. ***Right angles*** measure 90°. *Right angles theorem*: Any two right angles are congruent.

• Angles smaller than 90° are called ***acute angles***.

• Angles larger than 90° but less than 180° are called ***obtuse angles***.

• If the sum of any two angles is 180°, the two angles are called ***supplementary angles***. Supplementary angles sum to 180°. *Supplementary to same angle theorem*: If two angles are supplementary to the same angle, (or to congruent angles), then they are congruent to each other.

• If the sum of any two angles equals 90°, the two angles are called ***complementary angles***. Complementary angles sum to 90°. *Complementary to same angle theorem*: If two angles are complementary to the same angle (or to congruent angles), then they are congruent to each other.

• ***Angle addition postulate:*** If point B (on ray VB) is in between point A and point C on the interior of $\angle$AVC, then:

$m\angle AVB + m\angle BVC = m\angle AVC$

• ***Angle bisector theorem:*** A ray that bisects an angle divides it into two angles half the measure of the angle. If ray *BX* is the bisector of $\angle$ABC, then:

$m\angle ABX = (1/2)m\angle ABC$ and
$m\angle XBC = (1/2)m\angle ABC$

Notes

Chapter

3

Ratios and Proportions

One of the most famous proportions that has its origins in geometry is the *divine proportion*, also referred to as the *golden ratio* or *golden section*, *sectio aura*, and the *extreme and mean division*. The *golden ratio* is 1.618... and has interesting geometric properties, such as the aesthetically pleasing *golden rectangle*, whose length to width is the golden ratio, or the *golden triangle*, which is an isosceles triangle having its side in golden ratio to its base and if its base angle is bisected, a smaller golden triangle is formed. The golden ratio also receives attention and interest because of its manifestation in nature, art, and architecture. For example, the growth of spiral shells of mollusks, such as the chambered nautilus, obeys a pattern governed by the golden ratio. These shells have inspired many architectural constructions, such as Frank Lloyd Wright's Guggenheim Museum in New York City.

The golden section can be illustrated by drawing a line segment *AB* and dividing it by point C in such a way that *the whole segment (AB) is to the large part (AC) as the large part (AC) is to the small part (CB)*. Euclid described this as: *A straight line is said to have been cut in extreme and mean ratio when, as the whole line is to the greater segment, so is the greater to the lesser*.

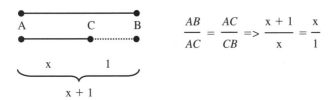

The *golden ratio* is derived by setting segment $CB = 1$ and $AC = x$, so that $AB = x + 1$. Next, substitute into ratio $AB/AC = AC/CB$ resulting in $(x + 1)/x = x/1$. Then, cross multiply: $x^2 = 1(x + 1) = x + 1$. Rearrange into the standard form of a *quadratic equation*, $ax^2 + bx + c = 0$, resulting in $x^2 - x - 1 = 0$. Solve using the **quadratic formula.**

$x = (-b \pm \sqrt{b^2 - 4ac})/2a$ and the values $a = 1$, $b = -1$, and $c = -1$:

$x = \left(-(-1) \pm \sqrt{-1^2 - 4(1)(-1)}\right)/2(1) = (1 \pm \sqrt{5})/2$

The positive root $(1 + \sqrt{5})/2$ is approximately 1.618, the **golden ratio!** This number is represented by the Greek letter phi (Φ) and is an irrational number, having decimal digits that never end and never repeat. Interesting *mathematical properties* of the golden ratio include:

$\Phi^2 = 2.618... = \Phi + 1$ and $1/\Phi = 0.618... = \Phi - 1$.

The **golden ratio** Φ is also associated with the **Fibonacci numbers**, named after the 12th-century Italian mathematician who first introduced them. The Fibonacci numbers are the sequence: 1, 1, 2, 3, 5, 8, 13, 21, 34, 55, ..., which is generated by beginning with 1 plus 1, and letting each subsequent number be the sum of its two predecessors so that the sequence is comprised of adding the previous two numbers together. If we take any Fibonacci number and divide it by its predecessor, the ratio will tend to the value of Φ, the golden ratio, as values become larger. For example, $55/34 \approx 89/55 \approx 233/144 \approx 1.618$. Also, as the series proceeds, any given number is 1.618 times the preceding number and 0.618% of the next number.

Fibonacci numbers, and by association the *golden ratio* Φ, are seen in various growth patterns in nature including the spiral arrangement of the kernels of a sunflower, the leaf arrangements of a tree, and the hexagonal scales on a pineapple. Properties of the Fibonacci series occur throughout nature, science, mathematics, music, and even business. In business, the Fibonacci related numbers 0.618, $0.382 = 0.618 \times 0.618$, and $0.5 = (0.382 + 0.618)/2$, are used in Technical Analysis of stock and commodity prices to determine potential support, resistance, and price levels. There are 38.2% retracements that imply a trend will continue, 61.8% retracements that imply a new trend is establishing itself, and 50% retracements that imply indecision.

3.1. Ratios and Proportions

Ratios

• A *ratio* represents a comparison between two numbers or quantities. For example, if the ratio between apples and oranges in a fruit bowl is 3 to 2, then for every 3 apples, there are 2 oranges. Ratios of two quantities can be represented as the first divided by the second so that the values are expressed in the form of a fraction or division:

3 apples / 2 oranges, or the ratio of apples to oranges is 3/2.

Ratios can also be thought of as the relationship between two similar values with respect to the number of times the first contains the second. In the ratio 6 feet / 3 feet = 2, there are two 3's in 6. More generally, we can say the ratio of the numbers x to y is the number x/y, where y ≠ 0. Because division by zero is undefined, a ratio cannot have zero in the denominator of its fraction.

• A ratio does not have *units of measure*, and is consequently called *abstract*. However, each of the two numbers that make up the ratio may have units (i.e., 4 feet / 2 feet = 2). When taking the ratio of two numbers having units, make sure they have been converted to the same units before dividing them. The division will then cancel those units so that the resulting ratio will not have units.

For example, the ratio of 6 inches to 1 foot, should be converted to 6 inches and 12 inches: 6 inches/12 inches = 1/2, or a ratio of 1 to 2. (The inches/inches cancelled.)

• Ratios are written 3 to 2, 3/2, or 3:2 (where a ":" can be used to represent "to" or "/"). A ratio can also be written as a decimal or a percent, such as 3/2 = 1.5 = 150%.

When a ratio is expressed as a fraction, it is usually written in reduced form, such as 3/2 rather than 6/4.

• There can be three or more numbers in a ratio, such as the three sides or three angles of a triangle. For example, a triangle with angles 30°, 60°, and 90° has an angle ratio of 30:60:90.

If three or more ratios are equal, an extended *proportion* can be written:

a/b = c/d = e/f

(See discussion on *proportions* following *ratios*.)

• *Side note*: The *difference between ratios and fractions* is that ratios represent part-per-part and fractions represent part-per-whole. Suppose a fruit bowl has 6 apples and 4 oranges:

The *ratio* of apples to oranges is 6/4 = 3/2, or 3 apples / 2 oranges.

The *fraction* of apples or oranges in the bowl is found by first determining the whole—or total number of apples and oranges together.

The total number of apples and oranges is:

   6 apples + 4 oranges = 10 fruit

The *fraction* of apples is 6/10 = 3/5, or 3 apples per every 5 pieces of fruit.

The *fraction* of oranges is 4/10 = 2/5, or 2 oranges per every 5 pieces of fruit.

They both use a fraction, but the meaning is different.

(See *Master Math: Basic Math and Pre-Algebra*.)

• **Example:** What is the ratio of: (a) 2(1/2) : (1/2), (b) 50¢ to $2.00, (c) 3x to x^2, and (d) 1 gal to 2 pt to 1 qt?

(a) First convert 2(1/2) = 5/2. Then, the ratio is: (5/2)/(1/2) = 5/1 = 5:1.
(b) The ratio: 50¢ to $2.00 = 50¢ to 200¢ = 50/200 = 5/20 = 1/4 = 1:4.
(c) The ratio: 3x to x^2 = $3x/x^2$ = 3/x = 3:x.
(d) The ratio: 1 gal to 2 pt to 1 qt = 4 qt to 1 qt to 1 qt = 4:1:1.

• **Example:** If the ratio of two complementary angles is 1:3, what is the measure of each angle?

The word *complementary* specifies that the angles sum to 90°:

   $m\angle a + m\angle b = 90°$

We are given that the ratio of the two angles is 1/3, so we can use a common value x that will multiply to equal each angle measure:

   1x/3x = 1/3

Therefore, we can write: $m\angle a + m\angle b = (1)x + (3)x = 90°$

Solve for x by first factoring:

   x(1 + 3) = 90°
   4x = 90°

Divide both sides by 4:

$x = 90°/4 = 22.5°$

Substitute x into the equation to find the value of the complementary angles:

$m\angle a + m\angle b = (1)x + (3)x = 90°$
$m\angle a + m\angle b = (1)22.5° + (3)22.5° = 90°$
$m\angle a + m\angle b = 22.5° + 67.5° = 90°$

Therefore, $m\angle a = 22.5°$ and $m\angle b = 67.5°$.

• **Example:** If the ratio of three angles is 2:1:3 and the second and third angles are supplementary, what is the measure of each angle?

There is a common value x that multiplies to equal each angle measure, so that we can write the angles as: 2x, 1x, and 3x.

The word *supplementary* means the angles sum to 180°, therefore the second and third angles are: $1x + 3x = 180°$. Solve for x:

$1x + 3x = 180°$
$4x = 180°$
$x = 180°/4 = 45°$
$x = 45°, 2x = 2(45°) = 90°, \text{ and } 3x = 3(45°) = 135°$

Therefore, the angles 2x, 1x, and 3x are 90°, 45°, and 135°, respectively.

• **Example:** If the sides of a triangle are in a ratio of 3:4:5 and the perimeter of the triangle is 36 miles, what is the measure of each side?

There is some common factor x for each side that will make the sum of the sides equal to the perimeter:

$3x + 4x + 5x = 36$

Solve for x by first factoring:

$x(3 + 4 + 5) = 36$
$12x = 36$

Divide both sides by 12:

$x = 36/12 = 3$

Substitute back into the equation to verify that the sum of the sides is the perimeter:

$$3x + 4x + 5x = 3(3) + 4(3) + 5(3) = 9 + 12 + 15 = 36$$

Therefore, the sides are: 9 miles, 12 miles, and 15 miles.

• **Example:** In an isosceles trapezoid, the bases AB and CD are parallel, the legs AD and BC are congruent, and the base angles ($\angle$A with $\angle$B and $\angle$C with $\angle$D) are congruent. In isosceles trapezoid ABCD, find: (a) the ratio of AB to CD, (b) the ratio of AB to BC, (c) the ratio $m\angle$A : $m\angle$B, (d) $m\angle$D / $m\angle$B, and (e) the ratio of BC to the perimeter of ABCD.

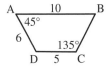

(a) The ratio of AB to CD is 10:5 = 2:1 = 2/1.
(b) The ratio of AB to BC is 10:6 = 5:3 = 5/3.
     (In an isosceles trapezoid the legs are congruent so $AD \cong$ BC.)
(c) $m\angle$A : $m\angle$B is 45:45 = 1:1 = 1/1.
     (In an isosceles trapezoid, the base angles are congruent so $m\angle$A = $m\angle$B and $m\angle$C = $m\angle$D.)
(d) $m\angle$D / $m\angle$B is 135:45 = 3:1 = 3/1.
(e) The perimeter of ABCD is the sum of its sides: 10 + 6 + 5 + 6 = 27. The ratio of BC to the perimeter of ABCD is 6:27 = 6/27.

Proportions

• *Mathematical **proportions** represent two equal ratios. Proportions are an equality between ratios.* Proportions are a comparative relation between the size or quantity of objects or numbers. The equation representing a proportion states that two ratios are equal.

• Proportions can be written in four equivalent forms:

1/2 is proportional to 2/4, 1/2 = 2/4, 1:2 = 2:4, 1:2::2:4

3/4 is proportional to 6/8, 3/4 = 6/8, 3:4 = 6:8, 3:4::6:8

a/b is proportional to c/d, a/b = c/d, a:b = c:d, a:b::c:d

• In the proportion a/b = c/d, the numbers represented by a, b, c, and d are referred to as the first, second, third, and fourth terms of the proportion, respectively. In a proportion, the middle numbers (second and third terms) are called the **means** of the proportion, and the outer numbers (first and fourth terms) are called the **extremes** of the proportion. In the proportion 1:2::2:4, the numbers 2 and 2 are the *means* and 1 and 4 are the *extremes*. The fourth term in a proportion is referred to as the **fourth proportional** to the other three terms taken in order. For example, in the proportion 1:2::2:4, we have 4 as the fourth proportional to 1, 2, and 2.

- If a/b is proportional to c/d, we can write:

  a/b̲ = c̲/**d**, **a**:b̲ = c̲:**d**, or **a**:b̲::c̲:**d**

  b and c (underlined) are the *means*.

  a and d (boldface) are the *extremes*.

  d is the *fourth proportional* to a, b, and c.

The following equation represents the proportion with means and extremes circled:

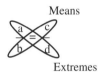

Means

Extremes

*When the means and extremes are **cross-multiplied**, the products are equal.*

- From algebra, we know that the equation 1/2 = 2/4 can be *cross-multiplied*, which is to multiply the opposite numerators with the opposite denominators of each fraction. Cross-multiplying results in an equivalent form of the same equation.

Cross-multiplying 1/2 = 2/4 results in:

  (1)(4) = (2)(2), which is equivalent to 4 = 4.

This principle is useful if we want to determine the value of one of the terms in a proportion (such as the side of a pair of similar triangles) when the ratio is known.

- We can demonstrate that cross multiplication is valid using a different approach by starting with $\dfrac{a}{b} = \dfrac{c}{d}$ and multiplying both sides by the common denominator bd: $\dfrac{abd}{b} = \dfrac{cbd}{d}$, then reduce each fraction where b/b = 1 and d/d = 1, resulting in ad = bc.

• **Example:** Find the fourth proportional of 3, 6, 9.

Write the proportion and solve for x:

3:6 = 9:x is equivalent to 3/6 = 9/x

Cross-multiply:

3x = (6)(9)
3x = 54
x = 54/3 = 18, where 18 is the fourth proportional.

• **Example:** Find the fourth proportional of a, b, c.

Write the proportion and solve for x:

a/b = c/x

Cross-multiply:

ax = bc
x = bc/a, where x is the fourth proportional.

Geometric Mean/Mean Proportional

• *If the two means terms of a proportion are the same,* then either *mean* is the ***mean proportional,*** also called the ***geometric mean,*** between the first and fourth terms.

For example, in proportion 1:2::2:4, the number 2 is the geometric mean between 1 and 4. We can also write this proportion as 1/2 = 2/4, which is equivalent (by cross-multiplying) to:

$$2^2 = (1)(4), \text{ or } 2 = \sqrt{(1)(4)}$$

This says that the geometric mean (mean proportional) between 1 and 4 is the square root of 1 times 4, which equals 2.

Similarly, in the proportion a:b::b:c, b is the geometric mean (mean proportional) between a and c.

• *Definition:* The number b is the ***geometric mean*** between numbers a and c if and only if b and c are greater than zero (because they are denominators and must not be zero in order to be defined), and the following equation is true: a/b = b/c.

In the proportion a:b::b:c, if b is the mean proportional, or geometric mean, then:

a/b = b/c

Cross-multiply:

$b^2 = ac$

b = square root (ac)

In other words, the geometric mean of a and c is the square root of their product. Or more generally:

The geometric mean of two numbers is the square root of their product.

• **Example:** What is the geometric mean, GM, of 4 and 16?

The geometric mean GM of 4 and 16 is the square root of their product:

GM(4,16) = square root(4×16) = square root(64) = 8

We can show this in the form of a proportion and cross-multiply:

$4/x = x/16$

$x^2 = (4)(16)$

$x = \sqrt{(4)(16)} = \sqrt{(2)(2)(4)(4)} = (2)(4) = 8$

• **Example:** What is the geometric mean between 8 and 10?

We can write the proportion and cross-multiply:

$8/x = x/10$

$x^2 = (8)(10) = 80$

$x = \sqrt{80} = \sqrt{(4)(4)(5)} = 4\sqrt{5}$

The geometric mean between 8 and 10 is $4\sqrt{5}$.

• *Side note:* The geometric mean should not be confused with the arithmetic mean. Remember, the *arithmetic mean* is the *average* of a group of numbers. The average is the sum of two or more numbers divided by how many numbers are added. For example, the arithmetic mean, or average of 2 + 3 + 4 is: (2 + 3 + 4)/3 = 9/3 = 3.

Properties of Proportions

• *Means-extremes/cross product property of a proportion:*

In any proportion, the product of the means equals the product of the extremes.

In addition, *if the product of the means equals the product of the extremes, then these pairs of numbers form a proportion.*

For example, 1:2::2:4 is a proportion and can therefore be written in the form: 1/2 = 2/4

This is equivalent to: (1)(4) = (2)(2) in which (1)(4) is the extremes and (2)(2) is the means.

Similarly, if we see (1)(4) = (2)(2), we know it can be algebraically rearranged as: 1/2 = 2/4.

The equation (1)(4) = (2)(2) can also be algebraically rearranged as:

    4/2 = 2/1

Therefore, using algebra, the proportion 1:2::2:4 can be rearranged and written in the following forms:

$$\frac{1}{2} = \frac{2}{4} \text{ is equivalent to (1)(4) = (2)(2) is equivalent to } \frac{4}{2} = \frac{2}{1}$$

which is equivalent to 4 = 4.

More generally, the equation (a)(d) = (b)(c) can be rearranged into the following four forms of proportions using simple algebra and multiplying and dividing both sides of the equation by a, b, c, or d:

$$\frac{a}{b} = \frac{c}{d} \qquad\qquad \frac{a}{c} = \frac{b}{d}$$

$$(a)(d) = (b)(c)$$

$$\frac{d}{b} = \frac{c}{a} \qquad\qquad \frac{d}{c} = \frac{b}{a}$$

Following are the steps of one algebraic rearrangement:

If we begin with the equation: (a)(d) = (b)(c)

Divide both sides by d:

    ad/d = bc/d

Because d/d is 1:

    a = bc/d

Divide both sides by c:

    a/c = bc/cd

Because c/c is 1:

    a/c = b/d

• These algebraic rearrangements of proportions are referred to as *proper-ties of proportions*. These properties include:

• The above stated *means-extremes/cross product property of a proportion:*

In a proportion, the product of the means equals the product of the extremes. The proportion:

$$\frac{a}{b} = \frac{c}{d} \text{ is equivalent to (a)(d)} = \text{(b)(c) is equivalent to } \frac{d}{c} = \frac{b}{a}$$

• ***The means or extremes switching property of a proportion:***

A proportion can be changed into an equivalent proportion by interchanging the means or the extremes.

$$\text{If } \frac{a}{b} = \frac{c}{d} \text{ is a proportion, } \frac{d}{b} = \frac{c}{a} \text{ and } \frac{a}{c} = \frac{b}{d} \text{ are equivalent}$$

• ***The inversion property of a proportion:***

A proportion can be changed into an equivalent proportion by inverting each ratio.

$$\text{If } \frac{a}{b} = \frac{c}{d} \text{ is a proportion, then } \frac{b}{a} = \frac{d}{c}$$

• ***The addition/subtraction property of a proportion:***

A proportion can be changed into an equivalent proportion by adding (or subtracting) terms in each ratio to obtain new first and third terms.

Addition property:

If a/b = c/d is a proportion, then (a + b)/b = (c + d)/d

Prove this property:

Given: a/b = c/d.

Prove: (a + b)/b = (c + d)/d.

Proof:

Statements	Reasons
1. a/b = c/d.	1. Given.
2. (a/b) + 1 = (c/d) + 1.	2. Addition: Add 1 to each side.
3. a/b + b/b = c/d + d/d.	3. Substitution: b/b = 1 and d/d = 1.
4. (a + b)/b = (c + d)/d.	4. Combine terms on each side with common denominator.

Example: Apply the addition property to $(a - 3)/3 = 6/2$:

$((a - 3) + 3)/3 = (6 + 2)/2$, or equivalently, $a/3 = 8/2$,
which is a less complicated form of the proportion.

Subtraction property:

If $a/b = c/d$ is a proportion, then $(a - b)/b = (c - d)/d$

Example: Apply the subtraction property to $(a + 3)/3 = 6/2$:

$((a + 3) - 3)/3 = (6 - 2)/2$, or equivalently, $a/3 = 4/2$,
which is a less complicated form of the proportion.

• *The corresponding terms property of a proportion*:

If any three terms in one proportion are equal to the corresponding three terms in a second proportion, then the remaining corresponding terms are equal.

If $2/3 = 4/x$ and $2/3 = 4/y$ are each proportions, then $x = y$. We can see this by solving for x and y: $x = (3)(4)/2 = 6$ and $y = (3)(4)/2 = 6$.

• *The corresponding sums property of a proportion*:

In a series of equivalent ratios, the sum of any of the numerators is to the sum of the corresponding denominators as any numerator is to its denominator.

If $a/b = c/d = e/f = ...$, then $(a + c + e + ...)/(b + d + f + ...) = a/b$

Example: If $2/3 = 4/6$, then $(2 + 4)/(3 + 6) = 6/9 = 2/3 = 4/6$

Example: If $2/3 = 4/6 = 6/9$, then $(2 + 4 + 6)/(3 + 6 + 9) = 12/18 = 6/9 = 2/3$

The corresponding sums property can be proven as follows:

Let j/k be the last term

$$\frac{a + c + e + \cdots + j}{b + d + f + \cdots + k}$$

Let $a/b = r$, $c/d = r$, $e/f = r$, and $j/k = r$, resulting in $a = br$, $c = dr$, $e = fr$, and $j = kr$:

$$\frac{br + dr + fr + \cdots + kr}{b + d + f + \cdots + k} = \frac{r(b + d + f + \cdots + k)}{b + d + f + \cdots + k}$$

$= r$ and $r = a/b$ therefore, $(a + c + e + ...)/(b + d + f + ...) = a/b$

• **Example:** We can *find a missing part of a proportion* using the *means-extremes/cross product property*. In the proportion $2:5::8:x$, find the missing part of the proportion.

We can find the missing part by solving for x:

2:5::8:x is equivalent to 2/5 = 8/x

Cross-multiply means and extremes:

2x = (5)(8)
2x = 40
x = 40/2 = 20

Therefore, the proportion is 2:5::8:20 and has 20 as its missing part.

- **Example:** Show that a/b = c/d is equivalent to (a + b)/b = (c + d)/d.

Use the means-extremes property and cross-multiply each equation:

Cross-multiply a/b = c/d: ad = bc
Cross-multiply (a + b)/b = (c + d)/d:
d(a + b) = b(c + d)
ad + bd = bc + bd

Subtract bd from each side: ad = bc

Cross-multiplying each equation, a/b = c/d and (a + b)/b = (c + d)/d, resulted in ad = bc. Therefore, they are equivalent to each other.

3.2. Proportional Segments

- If two segments (*AC* and *AD* below) are divided proportionally (by drawing *BE* parallel to *CD*), then resulting corresponding segments are proportional (their ratios are equal) to each other and to the original segments.

Segments *AC* and *AD* are divided proportionally, and the resulting segments, *AB* and *BC*, as well as *AE* and *ED*, are proportional to each other and to original segments *AC* and *AD*, so that:

AC/AB is proportional to *AD/AE*: *AC/AB* = *AD/AE*
AC/BC is proportional to *AD/ED*: *AC/BC* = *AD/ED*
BC/AB is proportional to *ED/AE*: *BC/AB* = *ED/AE*

Corresponding ratios are equal.

• *Side note*: In the above drawing, two *similar triangles* ΔADC and ΔAEB are formed. Two similar triangles can be created by drawing a line *parallel* to one of the sides of a triangle that intersects the other two sides. In the figure, ΔADC is *similar* to ΔAEB because the three corresponding angles are equal or congruent: ∠CAD ≅ ∠BAE, ∠ACD ≅ ∠ABE, and ∠ADC ≅ ∠AEB. (See Section 4.6. *Similar Triangles*.)

• In the previous section, we showed that the equation (a)(d) = (b)(c) can be rearranged using algebra by multiplying and dividing both sides of the equation by a, b, c, or d, into equivalent forms of proportions:

$$\frac{a}{b} = \frac{c}{d} \qquad\qquad \frac{a}{c} = \frac{b}{d}$$

$$(a)(d) = (b)(c)$$

$$\frac{d}{b} = \frac{c}{a} \qquad\qquad \frac{d}{c} = \frac{b}{a}$$

These proportions can represent proportionally-divided segments, and they correspond to the following four figures and eight proportion equations:

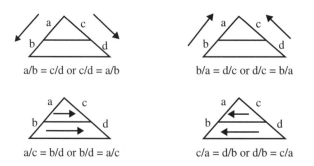

a/b = c/d or c/d = a/b   b/a = d/c or d/c = b/a

a/c = b/d or b/d = a/c   c/a = d/b or d/b = c/a

Notice that these figures representing proportionally-divided segments are triangles with a segment drawn parallel to one side. The following theorems apply to the triangle.

• ***Triangle proportionality theorem*** also called the ***side-splitter theorem***: If a line is parallel to one side of a triangle and intersects the other two sides, then it divides the other two sides proportionally.

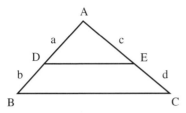

If $DE \parallel BC$, then a/b = c/d.
If a/b = c/d, then $DE \parallel BC$.
If $DE \parallel BC$, then a/AB = c/AC
and b/AB = d/AC.

Converse of triangle proportionality theorem: If a line divides two sides of a triangle proportionally, then it is parallel to the third side.

Corollary to triangle proportionality theorem: If a line is parallel to one side of a triangle and intersects the other two sides, then it produces segments proportional to the side lengths.

If $DE \parallel BC$, then a/AB = c/AC and b/AB = d/AC.

• **Example:** Prove the *triangle proportionality theorem,* which states that if a line is parallel to one side of a triangle and intersects the other two sides, then it divides the other two sides proportionally.

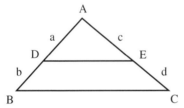

If $DE \parallel BC$, then a/b = c/d.

Given: ΔABC having $DE \parallel BC$.

Prove: a/b = c/d.

Plan: This proof uses the similarity of ΔABC and ΔADE in which ΔADE is created by DE being parallel to BC. In the proof show that ΔADE ~ ΔABC. Then show that AB/a = AC/c and by substitution is equivalent to b/a = d/c, which by the inversion property (or algebra) is equivalent to a/b = c/d. (This proof uses principles of similar triangles discussed in Chapter 4, Section 4.6. *Similar Triangles.*)

Proof:

Statements	Reasons
1. $\triangle ABC$ with $DE \parallel BC$.	1. Given.
2. $\angle ADE \cong \angle ABC$; $\angle AED \cong \angle ACB$.	2. If two parallel lines intersected by transversal corresponding $\angle$s are $\cong$.
3. $\triangle ADE \sim \triangle ABC$.	3. AA similarity post: If two $\angle$s of one $\triangle$ are $\cong$ two $\angle$s of other $\triangle$, then $\triangle$s are similar.
4. $AB/a = AC/c$.	4. Corresponding sides of similar triangles are in proportion.
5. $a + b = AB$; $c + d = AC$.	5. Segment addition postulate.
6. $(a + b)/a = (c + d)/c$.	6. Substitution of #5 into #4.
7. $b/a = d/c$.	7. Addition property of a proportion. $(a + b)/a = (c + d)/c => 1 + b/a = 1 + d/c => b/a = d/c$.
8. $b/a = d/c$ equals $a/b = c/d$.	8. Inversion property of proportion.

• **Example:** In the triangle $\triangle ABC$, if $a = 10$, $b = 6$, $c = 12$, and $DE \parallel BC$, find d.

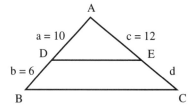

If $DE \parallel BC$, then $a/b = c/d$.

We can use the triangle proportionality theorem: If $DE \parallel BC$, then $a/b = c/d$. Then substitute given values into the proportion for a, b, and c, and solve for d:

$10/6 = 12/d$

Cross-multiply: $10d = 72$

$d = 7.2$

Therefore, $d = 7.2$.

• **Example:** In triangle $\triangle ABC$, if $DE \parallel BC$, $FE \parallel DC$, $AF = 2$, and $FD = 3$, find DB.

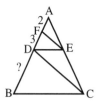

We can use the *triangle proportionality theorem*: If a line is parallel to one side of a triangle and intersects the other two sides, then it divides the other two sides proportionally.

In $\triangle ABC$, because $DE \parallel BC$, then $AD/DB = AE/EC$, or $5/DB = AE/EC$

In $\triangle ADC$, because $FE \parallel DC$, then $AF/FD = AE/EC$, or $2/3 = AE/EC$

Combine the two proportions. Because the right sides of the proportion are equal, $AE/EC = AE/EC$, then, the left sides are equal: $5/DB = 2/3$

Solve for DB by first cross-multiplying:

$2DB = (5)(3) = 15$

$DB = 15/2 = 7.5$

Therefore, $DB = 7.5$.

• We have learned that parallel lines in a triangle divide the sides proportionally. In a similar manner, any *three parallel lines with two transversals* crossing them divide the transversals proportionally.

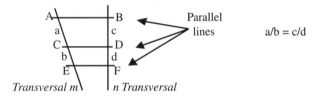

Three parallel lines transversals theorem: Three or more parallel lines divide any two transversals intersecting them proportionally.

In the figure a/b is proportional to c/d: $a/b = c/d$.

• **Example:** Prove the *three parallel lines transversals theorem.*

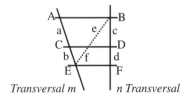

Given: $AB \parallel CD \parallel EF$.

Prove: a/b = c/d.

Plan: Create ΔEBF and ΔAEB. Use triangle proportionality theorem to show a/b = e/f and e/f = c/d, therefore by substitution a/b = c/d.

Proof:

Statements	Reasons
1. $AB \parallel CD \parallel EF$.	1. Given.
2. Draw BE across CD forming ΔEBF & ΔAEB. Label segments e & f.	2. Through 2 points exists one line.
3. a/b = e/f; e/f = c/d.	3. Δ proportionality theorem: If a line is $\parallel$ to one side of a Δ, then it divides the other two sides proportionally.
4. a/b = c/d.	4. Transitive axiom: Things equal to the same are equal to each other. (#3)

• **Example:** In the figure, if b = 3, c = 4, and c + d = 6, what is a?

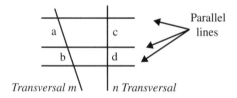

First find d using c + d = 6 and c = 4:

$$4 + d = 6$$
$$d = 6 - 4 = 2$$

We can use the *three parallel lines transversals theorem* and substitute into the proportion a/b = c/d (formed by the three parallel lines cut by two transversals):

$$a/3 = 4/2$$
$$2a = (4)(3)$$
$$a = 12/2 = 6$$

Therefore, a = 6 in the proportion a/b = c/d, or 6/3 = 4/2.

• If parallel lines are equidistant from each other, they will divide a transversal into congruent segments.

• *Three parallel lines congruent segments theorem*: If three parallel lines create congruent segments on one transversal, then they create congruent segments on every transversal.

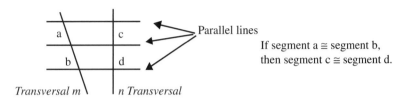

• **Example:** In the above figure, if a = 5, c = 4, and d = 4, what is b?

Using the *three parallel lines congruent segments theorem*, segment c ≅ segment d, therefore segment a ≅ segment b, and a = b = 5.

• A *bisector of an angle in a triangle* creates a proportion that can be used to determine side lengths.

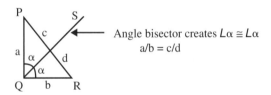

In ∆PQR angle bisector *QS* creates L𝛼 ≅ L𝛼 and proportion a/b = c/d.

• *Triangle angle bisector theorem*: A bisector of an angle of a *triangle* divides the opposite side (of the triangle) into segments, which are proportional to the adjacent sides.

In other words, if a ray or line bisects an angle of a triangle, then it divides the opposite side of the triangle into segments that are proportional to the other two sides.

In the above figure a/b is proportional to c/d, or a/b = c/d.

• **Example:** Prove the *triangle angle bisector theorem:*

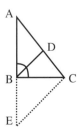

Given: △ABC with bisector *BD* of ∠ABC.

Prove: *AB/BC = AD/DC*.

Plan: Create △AEC with *CE* ‖ *BD*. By triangle proportionality theorem, *AB/BE = AD/DC*. Show that *BC* ≅ *BE* and substitute into proportion *AB/BE = AD/DC*. Use properties of parallel lines and transversals to show *BC* ≅ *BE* in isosceles △BCE.

Proof:

Statements	Reasons
1. Draw line *CE* ‖ *BD*.	1. Through a point not on a line, a line can be drawn parallel to that line.
2. Extend *AB* to intersect *CE* at E forming △AEC with *CE* ‖ *BD*.	2. Through 2 points exists 1 line. Also, two lines intersect at exactly 1 point.
3. *AB/BE = AD/DC*.	3. △ proportionality theorem: If a line is ‖ to one side of a △, it divides the other two sides proportionally.
4. *BD* bisects ∠ABC.	4. Given.
5. ∠ABD ≅ ∠DBC.	5. Definition of angle bisector.
6. ∠DBC ≅ ∠BCE.	6. If two ‖ lines (*CE* ‖ *BD*) are cut by transversal (*BC*), alternate interior angles are ≅.
7. ∠ABD ≅ ∠BCE.	7. Substitution #5 and #6.
8. ∠ABD ≅ ∠AEC.	8. If two ‖ lines (*CE* ‖ *BD*) are cut by transversal (*AE*), corresponding angles are ≅.
9. ∠AEC ≅ ∠BCE.	9. Substitution #7 and #8.
10. *BC* ≅ *BE* forming isosceles △BCE.	10. If 2 angles of a △ are ≅, then sides opposite those angles are ≅.
11. *AB/BC = AD/DC*.	11. Substitution #3 and #10.

• **Example:** In △PQR, if a = 120 inches, c = 8 feet, and d = 6 feet, find b.

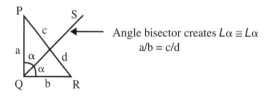

Angle bisector creates ∠α ≅ ∠α
a/b = c/d

First we must convert to the same units of measure. Let's convert a to feet: (120 inches)(1 foot/12 inches) = 10 feet (because inches/inches cancels).

Use the *triangle angle bisector theorem*: A bisector of an angle of a triangle divides the side opposite into segments proportional to the adjacent sides. In the proportion a/b = c/d, we can substitute given values and use algebra to solve for b:

   10/b = 8/6
   8b = (10)(6) = 60
   b = 60/8 = 7.5

Therefore, b = 7.5.

• **Example:** In the right triangle $\triangle ABC$, if leg *AB* = 9 miles, leg *BC* = 12 miles, and *BD* bisects angle B, find *AD* and *DC*. (This example uses the Pythagorean Theorem for right triangles discussed in Chapter 4, Section 4.8.)

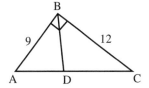

Using the *triangle angle bisector theorem*, which states that a bisector of an angle of a *triangle* divides the opposite side into segments which are proportional to the adjacent sides: *AB/BC = AD/DC*

   9/12 = *AD/DC*

Next, we can use the addition property of a proportion:

If a/b = c/d is a proportion, then (a + b)/b = (c + d)/d. Therefore:

   (9 + 12)/12 = (*AD + DC*)/*DC*

Notice that (*AD + DC*) is the hypotenuse of right $\triangle ABC$, which can be found using the Pythagorean Theorem $a^2 + b^2 = c^2$, or equivalently,

   $AB^2 + BC^2 = AC^2$: $9^2 + 12^2 = AC^2 = 81 + 144 = 225 = AC^2$
   $AC = \sqrt{225} = 15$
   $AC = (AD + DC) = 15$

Substitute back into (9 + 12)/12 = (*AD + DC*)/*DC*:

   21/12 = 15/*DC*
   DC(21) = (12)(15)
   DC = 180/21 ≈ 8.57

We can use the segment addition postulate $AD + DC = AC$ to find AD:

$AD = AC - DC = 15 - 180/21 = 315/21 - 180/21 = 135/21 \approx 6.43$

Therefore, given only the legs we have found $AD = 135/21 \approx 6.43$ and $DC = 180/21 \approx 8.57$.

To check we can substitute back into the proportion $AB/BC = AD/DC$:

$9/12 = (135/21) / (180/21) => 9/12 = 135/180$

$(180)(9) = (12)(135)$, or $1,620 = 1,620$

3.3 Chapter 3 Summary and Highlights

• A *ratio* represents a comparison between two numbers or quantities.

If the ratio of apples to oranges is 3 apples to 2 oranges, the ratio can be written 3 to 2, 3/2, 3:2, or as a decimal or a percent, $3/2 = 1.5 = 150\%$.

• *Mathematical **proportions** represent two equal ratios.* Proportions can be written: 1/2 is proportional to 2/4, $1/2 = 2/4$, 1:2 = 2:4, or 1:2::2:4.

• In a proportion, the middle numbers (second and third terms) are the *means*, and the outer numbers (first and fourth terms) are the *extremes*.

When the means and extremes are cross-multiplied, the products are equal.

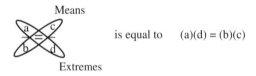

Means

is equal to   (a)(d) = (b)(c)

Extremes

• *If the two means terms of a proportion are the same*, then either *mean* is the *mean proportional*, also called the **geometric mean**, between the first and fourth terms. In the proportion a:b::b:c, if b is the geometric mean, then a/b = b/c is equal to $b^2 = ac$ is equal to $b = \sqrt{ac}$.

The geometric mean of two numbers is the square root of their product.

• A *proportion* can be arranged into the following equivalent forms:

a/b = c/d   a/c = b/d

(a)(d) = (b)(c)

d/b = c/a   d/c = b/a

• ***Triangle proportionality theorem***, also called the *side-splitter theorem*: If a line is parallel to one side of a triangle and intersects the other two sides, then it divides the other two sides proportionally.

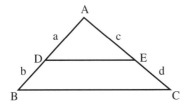

If *DE* ∥ *BC*, then a/b = c/d.
If a/b = c/d, then *DE* ∥ *BC*.
If *DE* ∥ *BC*, then a/*AB* = c/*AC*
and b/*AB* = d/*AC*.

• ***Triangle angle bisector theorem***: A bisector of an angle of a *triangle* divides the opposite side into segments which are proportional to the adjacent sides.

Notes

Chapter

4

Triangles, Congruence, and Similarity

Why are triangles so interesting and important? *Triangles* enable us to determine distances, heights, and angles, and are used in many fields, including surveying, astronomy, architecture, engineering, geography, and navigation. By drawing a model of a system and then drawing a triangle in the model with its sides, angles, and vertices representing key features, we can use the properties of triangles to calculate lengths, distances, and angle measurements. Properties of triangles are used not only in determining information about a triangle, but as we move on to other polygons and geometric figures, triangles can be drawn in to those figures and used to determine information. Properties of triangles are also used within proofs. The concepts of congruence and similarity are used throughout geometry and enable us to solve problems involving triangles and other geometric figures.

4.1 Triangle Definitions, Interior Angle Sum, and Exterior Angles

• *Triangles* are three-sided polygons that contain three angles. A triangle is formed by three line segments that join three *noncollinear* points. (Non-collinear points are points that are not in a straight line.) The three points where the three line segments that make up the triangle intersect or meet are called *vertices*. The line segments form the sides of the triangle. The *symbol for a triangle* is Δ. A triangle is named by the three letters at its vertices. For example, the triangle below is called ΔABC.

ΔABC has three vertices A, B, and C, three angles ∠A, ∠B, and ∠C, and three line segments *AB*, *BC*, and *AC*.

• **The *sum of the interior angles* in a planar triangle is always *180°*.**

This is the interior angles triangle sum theorem and is discussed on the following pages.

From this theorem, we find that if the value of two angles in a triangle is known, the third angle can be calculated by subtracting the sum of the two known angles from 180°.

• One important use of triangles is that a polygon in a plane having three or more sides can by subdivided into triangles so that triangle properties can be used to obtain information about the figure.

• Triangles can be *right* or *oblique*.

***Right triangles* are planar triangles with *one angle equal to 90°*.**

***Oblique triangles* are planar triangles that do *not have a 90° angle* and therefore are not right triangles.**

Oblique triangles may have all acute angles (<90°) or two acute angles and one obtuse angle (>90° but <180°). Right triangles can be isosceles, and oblique triangles can be equilateral or isosceles, which are discussed in Section 4.2. Both right triangles and oblique triangles can be used to model problems that require measurements of distances, lengths, and angles, such as determining the distance across a lake or canyon.

Right triangles: **Oblique triangles:**
One right angle.   No right angle.

Perimeter

- The *perimeter of a triangle* is the sum of its side lengths.

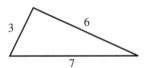

The perimeter of this triangle is: a + b + c

- **Example:** Find the perimeter of the triangle.

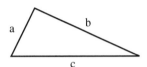

The perimeter is: $3 + 6 + 7 = 16$

Sum of the interior angles of a triangle

- *Interior angles triangle sum theorem*: **The sum of the interior angles in a planar triangle is always 180°.**

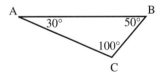

$m\angle A + m\angle B + m\angle C = 180°$

- We can also show that the sum of the angles in a triangle is 180° using the formula for the *sum of the interior angles of a polygon*, which is:

$(n - 2)180°$ with n equal to the number of vertices.

Because a triangle has three vertices then:

n = 3 and the formula becomes: $(3 - 2)180° = (1)180° = 180°$

• If you take the three angle measurements of any planar triangle and place them on a straight line, they will sum to 180°. In other words, *the sum of the measures of the angles in a triangle equals 180°, which is a straight angle.*

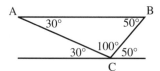

• We can prove the ***interior angles triangle sum theorem***; the sum of the interior angles in a planar triangle is always 180 degrees, as follows:

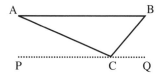

Given: ΔABC.

Prove: $m\angle A + m\angle B + m\angle C = 180°$.

Plan: Draw line PQ through vertex point C parallel to side AB. Because line PQ is a 180° angle, the sum of angles inside $\angle PCQ$ is $m\angle PCA + m\angle ACB + m\angle BCQ = 180°$. By the *parallel transversal postulate*, $\angle A \cong \angle PCA$ and $\angle B \cong \angle BCQ$, then substitute to obtain $m\angle A + m\angle B + m\angle C = 180°$.

Proof:

Statements	Reasons
1. Draw $PQ \parallel AB$ through C.	1. Parallel line postulate: A line can be drawn parallel to a given line through an external point.
2. $m\angle PCQ = 180°$.	2. A straight line angle measures 180°.
3. $m\angle PCA + m\angle ACB + m\angle BCQ = 180°$.	3. Partition axiom: Whole equals sum of parts.
4. $\angle A \cong \angle PCA$ and $\angle B \cong \angle BCQ$.	4. Parallel transversal postulate: Alternate interior angles of parallel lines $(PQ \parallel AB)$ cut by a transversal $(AC\ \&\ BC)$ are congruent.
5. $m\angle A + m\angle B + m\angle C = 180°$.	5. Substitution axiom: A quantity may be substituted for an equal quantity. (#3 & #4) $\angle C$ and $\angle ACB$ are the same angle.

Therefore, we have proved the theorem: The sum of the angles in a planar triangle is always 180 degrees.

• Following are *corollaries* of the *interior angles triangle sum theorem*, which are discussed throughout this chapter:

Corollary to triangle sum theorem: If the values of two angles in a triangle are known, the third angle can be calculated by subtracting the sum of the two known angles from 180°.

Corollary to triangle sum theorem: If two angles in one triangle are congruent to two angles in another triangle, the third angles will also be congruent.

Corollary to triangle sum theorem: Each angle in an equiangular triangle measures 60 degrees.

Corollary to triangle sum theorem: The two acute angles in a right triangle are complementary. (Complementary angles sum to 90°.)

Corollary to triangle sum theorem: There can be no more than one right or obtuse angle in a given triangle.

• **Example:** In the drawing below, the dashed line is parallel to the AB base of the $\triangle ABC$. Why is it true that $\angle A \cong \angle \theta$ and $\angle B \cong \angle \phi$?

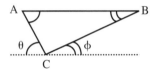

Because when two parallel lines (AB and line through C) are cut by a transversal (AC and BC), alternate interior angles are congruent.

• **Example:** Find the measure of $\angle C$.

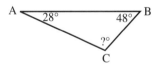

We can use the *interior angle triangle sum theorem*, which states that the sum of the interior angles in a planar triangle is always 180 degrees.

$$m\angle A + m\angle B + m\angle C = 180°$$

Substitute and solve for $m\angle C$:

$$28° + 48° + m\angle C = 180°$$
$$m\angle C = 180° - 28° - 48°$$
$$m\angle C = 104°$$

Check by substituting all three angle values:

$mLA + mLB + mLC = 180°$

$28° + 48° + 104° = 180°$, or $180° = 180°$

Therefore, $mLC = 104°$.

• **Example:** If Lc and Ld are congruent, are the triangles congruent?

We are given that $mLc = mLd$, and in addition, because vertical angles (La and Lb) are congruent, then $mLa = mLb$.

We can use the above corollary, *if two angles in one triangle are congruent to two angles in another triangle, the third angles will also be congruent*, to show that $mLe = mLf$.

Therefore, the three corresponding angles are congruent. However this does *not* necessarily mean that the triangles are congruent. If the three corresponding angles are congruent, then the triangles are called *similar*, which means they have the same shape, but may or may not have the same size. We were able to show that the three corresponding angles of the two triangles are congruent but this alone does not prove that the triangles are congruent. They may be congruent or just similar.

In Section 4.5. *Congruent Triangles* we learn that *to determine congruence for two triangles we need to identify at least three pairs of congruent parts which must include at least one pair of known side lengths.* Similar triangles are discussed in Section 4.6. *Similar Triangles* and 4.7. *Similar Right Triangles*.

Exterior Angles of a Triangle

• An *exterior angle of a triangle* (or any polygon) is an angle formed between the extension of one of the sides of the triangle and the outside of the triangle. Each exterior angle is a *supplement* of an adjacent interior angle so that an interior angle and its exterior angle sum to 180°. As a result, an exterior angle of a triangle forms a *linear pair* (180° angle) with one of the interior angles of the triangle.

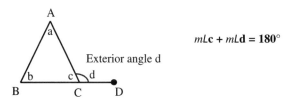

$$mL\mathbf{c} + mL\mathbf{d} = \mathbf{180°}$$

Note that Lc and Ld form a linear pair. Also, interior angles La and Lb are referred to as *remote, non-adjacent, or opposite* to exterior angle Ld.

- **Example:** What does a triangle look like if an exterior angle is acute?

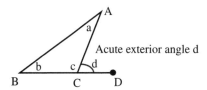

Acute exterior angle d

Because each exterior angle is a *supplement* of an adjacent interior angle, they sum to 180°: mLc $+ mL$d $= 180°$

If the exterior angle (Ld) is acute, and, by definition, measures less than 90°, then the adjacent interior angle (Lc) must measure greater than 90° and will by definition be an obtuse angle. If a triangle contains an obtuse angle, it is an *obtuse triangle*.

- An interesting property of triangles is the:

Exterior/remote interior angle theorem: **The measure of an exterior angle of a triangle is equal to the *sum* of the measures of the two non-adjacent opposite (remote) interior angles.**

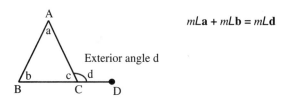

$$mL\mathbf{a} + mL\mathbf{b} = mL\mathbf{d}$$

Prove informally that the *exterior/remote interior angle theorem* is true.

Given: $\triangle$ABC with interior angles La, Lb, Lc, and exterior Ld.

Prove: mLa $+ mL$b $= mL$d.

Interior $\angle c$ is supplementary to exterior $\angle d$ and therefore:

$mLc + mLd = 180°$

The three interior angles of any planar triangle sum to 180°:

$mLa + mLb + mLc = 180°$

Combine the above equations, which both equal 180°:

$mLa + mLb + mLc = 180° = mLc + mLd$
$mLa + mLb + mLc = mLc + mLd$

Subtract mLc from both sides:

$mLa + mLb = mLd$

• From (preceding) *exterior/remote interior angle theorem*,
$La + Lb = Ld$, it would follow that Ld must be greater than either
La or Lb.

Exterior angle inequality theorem: **The measure of an exterior angle of
a triangle is *greater than* the measure of either non-adjacent remote
interior angle.**

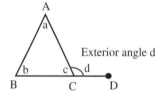

$mLd > mLa$ and $mLd > mLb$

Exterior angle d

• **Example:** Using the figure below: (a) Can mLd equal 53°? (b) Find
mLc and mLd.

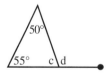

(a) Using the *exterior angle inequality theorem*, the measure of an exterior
angle of a triangle is *greater than* the measure of either non-adjacent
remote interior angle, we know that $mLd > 50°$ and $mLd > 55°$; therefore,
mLd must be greater than both 50° and 55° and cannot be 53°.

(b) Using *exterior/remote interior angle theorem*, an exterior angle of a triangle is equal to the sum of the two remote (non-adjacent) interior angles, we can calculate mLd:

$$50° + 55° = mLd = 105°$$

Because Lc and Ld are supplementary and therefore sum to 180°:

$$mLc + mLd = 180°$$

Substitute $mLd = 105°$ and subtract it from both sides:

$$mLc = 180° - 105° = 75°$$

Therefore, $mLc = 75°$ and $mLd = 105°$.

An alternative approach to solving part (b) is to find mLc using the *interior angles triangle sum theorem*, the sum of the angles in a planar triangle is always 180 degrees:

$$50° + 55° + mLc = 180°$$
$$105° + mLc = 180°$$

Isolate mLc by subtracting 105° from both sides of the equation:

$$mLc = 180° - 105° = 75°$$

Because Lc and Ld are supplementary and $mLc = 75°$:

$$mLc + mLd = 180°$$
$$75° + mLd = 180°$$

Subtract $= 75°$ from both sides:

$$mLd = 180° - 75° = 105°$$

Therefore, $mLc = 75°$ and $mLd = 105°$.

• In summary:

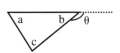

$$mL\theta = mLa + mLc$$
$$mLa + mLb + mLc = 180°$$
$$mLb + mL\theta = 180°$$

• **Exterior angles sum theorem**: The sum of the measures of the exterior angles of a triangle equals 360°.

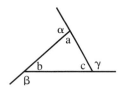

Interior angles: $mLa + mLb + mLc = 180°$
Exterior angles: $mL\alpha + mL\beta + mL\gamma = 360°$

Because each exterior angle is a supplement of an adjacent interior angle, if you take the supplements of the interior angles and sum them, you can prove to yourself that the exterior angles sum to 360°.

For example, if the interior angles are 80°, 40°, and 60°, then their supplements are 100°, 140°, and 120°, respectively.

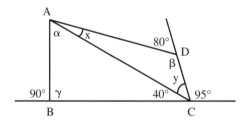

$$80° + \ 40° + \ \ 60° = 180° \ \text{Interior angles}$$
$$100° + 140° + \ 120° = 360° \ \text{Exterior angles}$$
$$\overline{180° \quad 180° \quad \ 180°}$$

• **Example:** (a) Find x and y. (b) Do the interior angles in quadrilateral ABCD sum to 360°?

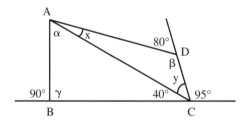

(a) To find angle y we know that $y + 40° + 95°$ must equal 180° because they lie on straight line BC. Therefore:

$$y + 40° + 95° = 180°$$
$$y = 180° - 40° - 95° = 45°$$
$$y = 45°$$

To find angle x, we use the *interior angles triangle sum theorem,* which states that the angles in a triangle sum to 180°. In triangle $\triangle ACD$:

$$x + y + \beta = 180°.$$

Angle β is the supplement of the exterior 80° angle, and because each exterior angle is a supplement of an adjacent interior angle and supplementary angles sum to 180°, then:

$$\beta + 80° = 180°$$
$$\beta = 180° - 80° = 100°$$

Substitute $\beta = 100°$ and $y = 45°$ into the triangle angle sum equation, $x + y + \beta = 180°$, and solve for x:

   $x + 45° + 100° = 180°$
   $x = 180° - 45° - 100° = 35°$
   $x = 35°$

Therefore, $x = 35°$ and $y = 45°$.

(b) The angles in quadrilateral ABCD are $(\alpha + x)$, (γ), $(40° + y)$, and (β).

   γ is the supplement of the 90° external angle and is therefore 90°.
   α is the third angle in triangle $\triangle ABC$: $\alpha + \gamma + 40° = 180°$, or
   $\alpha = 180° - 90° - 40° = 50°$.
   $\beta = 100°$ (β is supplement to external 80° angle)
   $x = 35°$ (from part (a))
   $y = 45°$ (from part (a))

Add up the interior angles in quadrilateral ABCD using the values for each angle:

   $(\alpha + x) + (\gamma) + (40° + y) + (\beta)$
   $= (85°) + (90°) + (85°) + (100°) = 360°$

The angles in quadrilateral ABCD correctly sum to 360°.

Note: We will learn in Chapter 5 that the sum of the angles in any planar quadrilateral is 360°.

4.2 Types of Triangles

• Triangles are classified by both their side lengths and by the type of angles they have.

Types of triangles include:

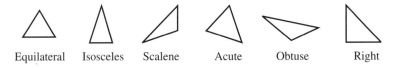

Equilateral   Isosceles   Scalene   Acute   Obtuse   Right

Equilateral Triangle

• In an *equilateral triangle*, *all three sides have equal lengths, and all three angles have equal measurements of 60°.*

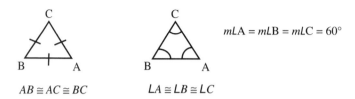

$mLA = mLB = mLC = 60°$

$AB \cong AC \cong BC$ $LA \cong LB \cong LC$

• *Equilateral* triangles are *equiangular*.

• *Equilateral/equiangular theorem*: If a triangle is equilateral, it is also equiangular. (See proof in example on page 108.)

• *Equiangular/equilateral theorem*: If a triangle is equiangular, it is also equilateral.

• *Equiangular 60° theorem*: Each angle in an *equiangular triangular* has a measure of 60°.

• An *equilateral triangle* is also a type of an *isosceles triangle*.

Isosceles Triangle

• In an *isosceles triangle*, two *sides* have equal lengths, and the *angles* opposite the two equal sides have equal measurements.

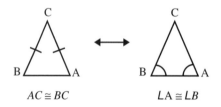

$AC \cong BC$ $LA \cong LB$

• *Congruent sides/angles theorem*: **If two sides of a triangle are congruent, then the angles opposite those sides are congruent.** (See proof in example page 106.)

• *Congruent angles/sides theorem*: **If two angles of a triangle are congruent, then the sides opposite those angles are congruent.**

• The parts of an *isosceles triangle*:

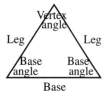

• *Side note*: Each of the four faces of the Transamerica Pyramid building in San Francisco is an isosceles triangle.

Scalene Triangle

• In a *scalene triangle*, all three *sides* have different lengths, and all three angles have different measurements.

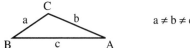

$a \neq b \neq c$ and $m\angle A \neq m\angle B \neq m\angle C$

Acute Triangle

• In an *acute triangle*, all three interior *angles* measure less than 90°.

$m\angle a < 90°$; $m\angle b < 90°$; $m\angle c < 90°$

Obtuse Triangle

• In an *obtuse triangle*, one of the interior *angles* is larger than 90° (but less than 180°).

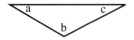

$m\angle a < 90°$; $m\angle b > 90°$; $m\angle c < 90°$

A triangle can have no more than one obtuse angle (because the total sum of the angles is 180° and if one angle is greater than 90°, the other two must be less than 90°).

Right Triangle

• In a *right triangle*, one of the interior *angles* in the triangle is a *right angle* measuring 90°.

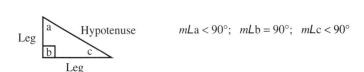

$m\angle a < 90°$;　$m\angle b = 90°$;　$m\angle c < 90°$

□ denotes the 90° angle. The hypotenuse is always the longest side because the 90° angle is the largest angle in a right triangle. This is because the sum of the angles in a planar triangle is always 180°, and so the remaining two angles must each measure less than 90°.

• Right triangles possess one right (90°) angle and two acute (<90°) angles that sum to 90°, so that the total sum of the angles is 180°.

• The *two acute angles* of a right triangle are called **complementary** and sum to 90°.

• **Example:** What type of triangle has the characteristic that the sum of the measures of two of its angles is equal to the measure of the third angle?

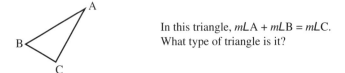

In this triangle, $m\angle A + m\angle B = m\angle C$.
What type of triangle is it?

The type of triangle that has this characteristic is a right triangle.

This is because the sum of the angles in a planar triangle is always 180°, and the right angle, by definition, measures 90°; therefore, the remaining two angles must sum to 90°.

• **Example:** In this *isosceles triangle* if the vertex angle $\angle c$ is 30°, what is the measure of the two congruent angles $\angle a$ and $\angle b$?

Because $mLa + mLb + mLc = 180°$ and $mLa = mLb$, then we can write:

$mLa + mLa + mLc = 180°$

Substitute $mLc = 30°$, and solve for mLa:

$2mLa + 30° = 180°$
$2mLa = 150°$
$mLa = 150°/2 = 75°$

Therefore, $mLa = mLb = 75°$.

To check your result substitute back into $mLa + mLb + mLc = 180°$:

$75° + 75° + 30° = 180°$

- **Example:** In the triangle $\triangle ABC$, which angles are congruent?

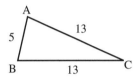

LA and LB are congruent because they are opposite congruent sides CA and CB.

- **Example:** In the figure below, which sides are congruent?

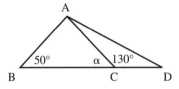

$L\alpha$ is a supplementary angle with the $130°$ angle; therefore,

$mL\alpha + 130° = 180°$, or
$mL\alpha = 180° - 130° = 50°$

Because $mL\alpha = 50°$ and $mLB = 50°$, and $L\alpha$ is congruent to LB, the sides opposite $L\alpha$ and LB are congruent. These congruent sides are AB and AC. Therefore, the congruent sides are AB and AC of $\triangle ABC$.

• **Example:** Find $\angle$a, $\angle$b, and $\angle$c. (Tick marks indicate congruency.)

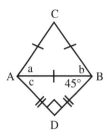

We can see from the small square at vertex D, the double tick marks, and the 45° angle that $\triangle$ABD is a right isosceles (45:45:90) triangle. Angles opposite congruent sides are congruent; therefore: $m\angle$c = 45°.

From the single tick marks, we can see that $\triangle$ABC is an equilateral triangle. Because equilateral triangles are also equiangular, and the *equiangular 60° theorem* states that each angle in an equiangular triangle has a measure of 60°, then: $m\angle$a = $m\angle$b = 60°.

Therefore, $m\angle$a = 60°, $m\angle$b = 60°, and $m\angle$c = 45°.

• **Example:** Prove *congruent sides/angles theorem*, which states that if two sides of a triangle are congruent, then the angles opposite those sides are congruent.

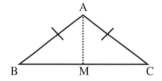

Given: $AB \cong AC$ ($\triangle$ABC is an isosceles triangle).

Prove: $m\angle$B = $m\angle$C, or $\angle$B $\cong \angle$C.

Plan: Draw angle bisector AM of $\angle$A. Show $\triangle$AMB $\cong \triangle$AMC by SAS. Because corresponding parts of congruent triangles are congruent, $m\angle$B = $m\angle$C. (This proof uses properties of *congruent triangles* discussed in Section 4.5.)

Proof:

Statements	Reasons
1. $AB \cong AC$.	1. Given/definition of isosceles triangle.
2. Draw AM bisector of LA.	2. An angle has one bisector.
3. $mL\text{BAM} = mL\text{CAM}$.	3. Definition of angle bisector.
4. $AM \cong AM$.	4. Identity or reflexive axiom: Self = self.
5. $\triangle\text{BAM} \cong \triangle\text{CAM}$.	5. SAS congruence postulate: If the two sides and included angle of one $\triangle$ are $\cong$ to corresponding parts of other $\triangle$, then the $\triangle$s are $\cong$. (#1, #3, & #4)
6. $LB \cong LC$.	6. Corresponding parts of $\cong$ $\triangle$s are $\cong$.

- **Example:** In the *isosceles triangle* $\triangle$ABC, with Z as a midpoint and $mL\alpha = mL\beta$ ($L\alpha \cong L\beta$), prove that XZ is congruent to YZ.

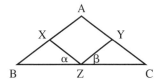

Given: $\triangle$ABC is an isosceles triangle; Z is the midpoint of BC; $mL\alpha = mL\beta$.

Prove: $XZ \cong YZ$.

Plan: Show $\triangle$BZX $\cong$ $\triangle$CZY using $L\alpha \cong L\beta$, $LB \cong LC$, and $BZ \cong ZC$. Because corresponding parts of congruent triangles are congruent, $XZ \cong YZ$. (This proof uses properties of *congruent triangles* discussed in Section 4.5.)

Proof:

Statements	Reasons
1. Isosc. $\triangle$ABC; Z midpoint of BC.	1. Given.
2. $AB \cong AC$.	2. Definition of isosceles triangle.
3. $L\alpha \cong L\beta$.	3. Given.
4. $BZ \cong ZC$.	4. Definition of midpoint.
5. $LB \cong LC$.	5. Angles opposite congruent sides of a triangle are congruent to each other.
6. $\triangle$BZX $\cong$ $\triangle$CZY.	6. ASA congruence postulate: If the 2 Ls and included side of one $\triangle$ are $\cong$ to corresponding parts of other $\triangle$, $\triangle$s are $\cong$.
7. $XZ \cong YZ$.	7. Corresponding parts of $\cong$ $\triangle$s are $\cong$.

• **Example:** Prove *equilateral/equiangular theorem*, which states that if a triangle is equilateral, it is also equiangular.

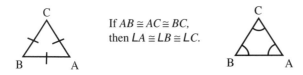

If $AB \cong AC \cong BC$,
then $\angle A \cong \angle B \cong \angle C$.

Given: $\triangle ABC$ is an equilateral triangle.

Prove: $\triangle ABC$ is an equiangular triangle.

Plan: Show all angles congruent using *congruent sides/angles theorem*: If two sides of a triangle are congruent, the angles opposite those sides are congruent (proved above).

Proof:

Statements	Reasons
1. $\triangle ABC$ is an equilateral triangle.	1. Given.
2. $AB \cong AC \cong BC$.	2. Definition of equilateral triangle.
3. $m\angle C = m\angle B$; $m\angle B = m\angle A$.	3. If two sides of a $\triangle$ are $\cong$, the angles opposite those sides are $\cong$.
4. $m\angle C = m\angle A$.	4. Transitive axiom: Things $\cong$ to same are $\cong$ to each other. (#3)
5. $\triangle ABC$ is an equiangular $\triangle$.	5. If all $\angle$s in a $\triangle$ are $\cong$, the $\triangle$ is equiangular. $m\angle C = m\angle B$, $m\angle B = m\angle A$, $m\angle C = m\angle A$.

4.3 Parts of Triangles, Altitude, Bisector, Median, and Ceva's Theorem

• Triangles generally consist of three points and three segments. There are also special names for these parts as well as names for additional points and segments that may be included in a triangle. Names or terms associated with triangles include: *vertex or vertices*, *angle*, *side*, *leg*, *base*, *altitude*, *bisector*, *midpoint*, *median*, *cevian*, and *hypotenuse*. These parts are identified in the following paragraphs.

• In a *right triangle*, the side opposite the right angle is called the *hypotenuse* and the other two sides adjacent to the right angle are called *legs*. (The legs in a right triangle may or may not have the same length.)

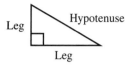

Right triangles contain
one right angle.

• In an *isosceles triangle*, the two equal-length sides are called *legs*, and the non-equal side is called the *base*. Also, the two congruent angles (opposite the two congruent sides) are called *base angles*, and the third, non-congruent angle is called the *vertex angle*. (Note: The term base is more generally used to describe any side of any triangle to which an altitude line is drawn. This is discussed with altitudes below.)

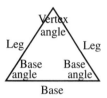

Isosceles triangles contain
two congruent sides (legs),
which are opposite two
congruent angles (base angles).

Midpoint

• The *midpoint* of a side of a triangle is the point equidistant from each vertex and in the center of the side.

• *Midpoint to midpoint theorem*: In a triangle, if a line is drawn from the *midpoint* of one side and parallel with a second side, then it will pass through the midpoint of the third side.

In $\triangle ABC$, if X is the midpoint of *AB* and
XY ∥ *BC*, then Y is the midpoint of *AC*.

• *Midpoints parallel length theorem*: **In a triangle, if a line connects the midpoints of two sides, then it will be parallel to the third side, and equal to half the length of the third side.**

In $\triangle ABC$, if X is the midpoint of *AB*
and Y is the midpoint of *AC*,
then *XY* ∥ *BC* and *XY* = (1/2)BC.

(See Section 4.6. *Similar Triangles*, page 150 for a proof.)

• *Right triangle midpoint theorem*: **In a right triangle, the midpoint of the hypotenuse is equidistant from the three vertices.**

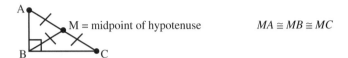

$MA \cong MB \cong MC$

Perpendicular Bisector and Midpoint

• **A *perpendicular bisector of the side* of a triangle is a line, ray, or segment that is perpendicular to and bisects that side at its *midpoint*.**

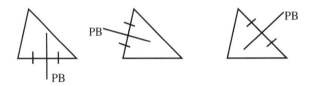

A perpendicular bisector is perpendicular to the side its bisects. Note: A median (described following) and a perpendicular bisector intersect the side at the same point.

• *Perpendicular bisectors concurrent theorem*: The *perpendicular bisectors of the sides of a triangle are concurrent and intersect* at a point that is *equidistant* from the three vertices of the triangle. (Concurrent lines intersect at a single point.)

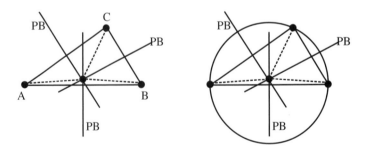

Note: The perpendicular bisectors of the sides of a *triangle* intersect at the center of a *circle* that circumscribes the triangle and has the distance from center to each vertex as its radius (dashed).

- Also, remember the *theorems* from Chapter 2 *Points, Lines, Planes, and Angles:*

Perpendicular bisector/endpoints theorem: If a point lies on the *perpendicular bisector* of a segment, then the point is equidistant from the *endpoints* of the segment.

Endpoints/perpendicular bisector theorem: If a point is equidistant from the *endpoints* of a segment, then it lies on the *perpendicular bisector* of the segment.

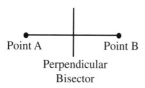

Median

- A *median* of a triangle is a line segment extended from any of the three angle vertices to the *center or midpoint* of the side opposite the angle. A median *bisects* the side to which it is drawn.

- Note: A median and a perpendicular bisector intersect the side at the same point, but the median is not necessarily perpendicular.

 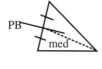

- The *three medians* drawn from the three angles of any triangle are **concurrent,** which means they meet at a single point *inside the triangle* (even obtuse triangles).

Medians concurrent theorem: The three *medians* of a triangle are concurrent.

(See discussion of Ceva's Theorem page 120 for a proof.)

Medians are concurrent

Obtuse triangle   Right triangle   Acute triangle

• **Centroid**: The *point in a triangle where the three **medians** intersect* each other (are concurrent) is the **center of gravity**, or **centroid**, of the triangle. This means that if a triangle was made of a solid material, it would *balance on its centroid* (where the three medians intersect).

• ***Centroid two-thirds theorem***: The point in a triangle where the three *medians* are concurrent (intersect each other) is *two-thirds* the distance from each vertex to the opposite side.

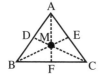

Medians intersect at point M, which is two-thirds distance from vertex to side.

Length *AM* equals 2/3 of the length of *AF*, or M is 2/3 distance from A to F.
Length *BM* equals 2/3 of the length of *BE*, or M is 2/3 distance from B to E.
Length *CM* equals 2/3 of the length of *CD*, or M is 2/3 distance from C to D.

• **Example:** In the above figure, if M is the point of concurrency and median *BE* is 24 meters, what is the length of *BM*?

Length *BM* equals 2/3 of the length of *BE*, or

$$BM = (2/3)BE$$
$$BM = (2/3)(24 \text{ meters}) = 16 \text{ meters}$$

Therefore, length *BM* equals 16 meters.

• ***Right triangle median theorem***: In a right triangle, the median from the right angle to the hypotenuse is one-half the length of the hypotenuse.

Median $AM = (1/2)BC$

(See example in Section 4.8 Right Triangles.)

• *Median triangle area theorem*: A median divides a triangle into two triangles having equal areas.

Area $\triangle ACM \cong$ Area $\triangle ABM$

Altitude and Base

• **An *altitude* of a triangle is the segment from a vertex perpendicular to the opposite side.** An *altitude* of a triangle can be drawn from any of the three angles, and is a line segment that extends from the angle *perpendicular* to the side opposite the angle. The side to which an altitude is drawn is then referred to as a ***base***. The altitude is perpendicular to the base and meets the base at a right angle. The term ***base*** can be applied to any side of any triangle as it refers to the *altitudes* of the triangle.

• *Altitude lines* drawn in an *acute triangle*:

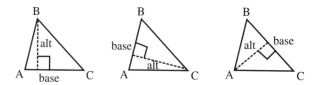

• *Altitude lines* drawn in a *right triangle* have two of the *altitudes* coinciding with one of the legs:

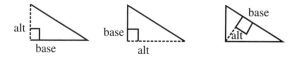

• *Altitude lines* drawn in an *obtuse triangle* have two of the *altitudes* extending outside the triangle:

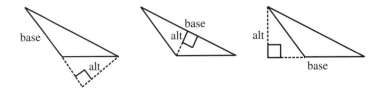

• Note: If the area of a triangle is known, the altitude of a triangle can be calculated using the triangle area formula:

   Area of a triangle = (1/2)(base)(altitude)

(See Section 4.10. *Area of a Triangle*.)

• The *three altitudes* drawn from the three angles of any triangle meet at a single point (are *concurrent*), which may or may not be inside the triangle.

Altitudes concurrent theorem: The three *altitude lines* of a triangle are concurrent. (See page 121.)

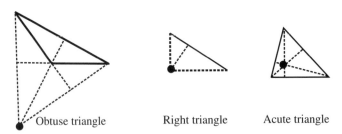

Obtuse triangle   Right triangle   Acute triangle

The points represent the orthocenters of the triangles.

• *Definition*: The **orthocenter** of a triangle is the point where the altitude lines are concurrent.

Angle Bisector

• **An *angle bisector* of a triangle is a line, segment, or ray that bisects the angle and is extended from any of the three angles to the side opposite the angle. An *angle bisector* divides the angle in half, resulting in two congruent angles which are half the measure of the vertex angle that is bisected.** (AB denotes angle bisector.)

The angle bisector divides the angle into two congruent angles.

Angle bisector to sides theorem: If a point is on the *bisector of an angle*, then it is *equidistant* from the sides of the angle.

Sides to angle bisector theorem: If a point is equidistant from the sides of an angle, then it is on the bisector of the angle.

• In Chapter 3, we learned that a ***bisector of an angle in a triangle*** creates a proportion that can be used to determine side lengths.

Triangle angle bisector theorem: A bisector of an angle of a *triangle* divides the opposite side (of the triangle) into segments, which are proportional to the adjacent sides.

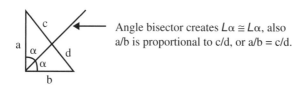

Angle bisector creates $L\alpha \cong L\alpha$, also a/b is proportional to c/d, or a/b = c/d.

See Chapter 3, Section 3.2. for proof and examples.

• The *three angle bisectors* drawn from the three angles of any triangle are ***concurrent***, and meet at a single point *inside the triangle* (even obtuse triangles). In addition, the point where the three angle bisectors intersect is the center of a circle that can be inscribed into the triangle. See Construction 23 in Chapter 8 *Constructions and Loci.*

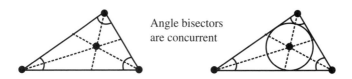

Angle bisectors are concurrent

Angle bisectors concurrent theorem: The interior angle bisectors of any triangle are concurrent. (See page 120.)

Equidistant angle bisectors theorem: The bisectors of the angles of a triangle are concurrent, and intersect at a point that is equidistant from the three sides of the triangle.

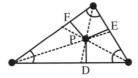

Angle bisector intersection point is equidistant from each side.

$PD \cong PE \cong PF$

Combining altitude, angle bisector, median, and perpendicular bisector

Altitude perpendicular to base.	*Angle bisector* bisects angle.	*Median* bisects base at midpoint.	*Perpendicular bisector* bisects side at midpoint.

• The *altitude* drawn from the vertex angle of an *isosceles triangle* or any angle of an *equilateral triangle* is also the *angle bisector*, as well as the *median*.

  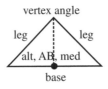

In an equilateral triangle or from the vertex angle of an isosceles triangle:

altitude = median = angle bisector

Equilateral triangle   Isosceles triangle

• **Example:** Prove that the *angle bisector* drawn from the vertex angle of an isosceles triangle (or any angle of equilateral triangle) is also the *median* of the base.

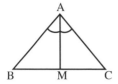

Given: $\triangle ABC$ is an isosceles triangle; therefore, $AB \cong AC$;

AM bisects ∠A.

Prove: Angle bisector AM is also the *median* of base BC.

Plan: Show $\triangle BAM \cong \triangle CAM$ using SAS congruence postulate with $AB \cong AC$, ∠BAM ≅ ∠CAM (AM bisects ∠A), and $AM \cong AM$. Because corresponding parts of congruent triangles are congruent, $BM \cong CM$, and AM is the median of BC. (This proof uses *congruent triangles* discussed in Section 4.5.)

Proof:

Statements	Reasons
1. $AB \cong AC$.	1. Given. Definition of isosceles triangle.
2. AM bisects ∠A.	2. Given.
3. ∠BAM ≅ ∠CAM.	3. Def. of ∠ bisector: Divide ∠ into ≅ ∠s.
4. $AM \cong AM$.	4. Identity or reflexive axiom: Any quantity is equal to itself.
5. $\triangle BAM \cong \triangle CAM$.	5. SAS congruence postulate: If 2 sides and included ∠ of one △ are ≅ to corresponding parts of other △, △s are ≅. (#1, #3, #4)
6. $BM \cong CM$.	6. Corresponding parts of ≅ △s are ≅.
7. AM is the median of BC.	7. Def. of median of a triangle: Line from angle vertex bisecting opposite side.

- **Example:** If $DE \parallel BC$ and $FE \parallel BA$, find x and y.

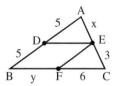

To solve this problem, we can use the *midpoint to midpoint theorem*, which states that in a triangle, if a line is drawn from the *midpoint* of one side and parallel with a second side, then it will pass through the midpoint of the third side. Because $DE \parallel BC$ and D is the midpoint of AB (5 units from A and B), then E is the midpoint of AC.

This results in x = 3.

Similarly, because $FE \parallel BA$ and E is the midpoint of AC (3 units from A and C), then F is the midpoint of BC.

This results in y = 6.

Therefore, x = 3 and y = 6.

Ceva's Theorem

• Any line segment that joins a vertex of a triangle to any given point on the opposite side is called a *cevian*. Altitudes, medians, and angle bisectors are all cevians. If L, M, and N are points on respective sides BC, AC, and AB of △ABC, then segments AL, BM, and CN are cevians.

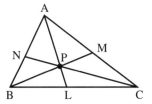

Ceva's Theorem: The three cevians AL, BM, and CN of △ABC, one through each vertex A, B, and C to points on the opposite sides L, M, and N, respectively, are *concurrent*, if and only if

$$(AN/NB)(BL/LC)(CM/MA) = 1$$

Ceva's Theorem is also written:

If three cevians AL, BM, and CN of △ABC, one through each vertex A, B, and C, are concurrent, then

$$(AN/NB)(BL/LC)(CM/MA) = 1$$

Note: When lines or segments are *concurrent*, they all pass through one point.

• Prove *Ceva's Theorem*:

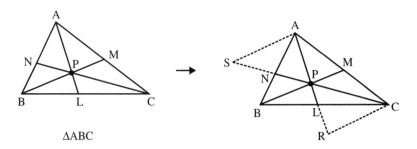

△ABC

There are several approaches that can be used to *prove Ceva's Theorem*, which can be found in various geometry books. One method involves the drawing above combined with the use of similar triangles and their proportional sides. (The symbol for *similar* is ~. See Section 4.6. for an explanation of proportional sides in similar triangles.)

In ∆ABC (on the left), we can construct the figure on the right by drawing a line through point A (*SA*) that is parallel to *BP* and then drawing a line through C (*CR*) that is parallel to *BP*.

This results in four useful pairs of similar triangles:

(a) ∆ASN ~ ∆BPN and their sides are in proportion: *AN/NB* = *AS/BP*

(b) ∆BPL ~ ∆CRL and their sides are in proportion: *BL/LC* = *BP/CR*

(c) ∆PAM ~ ∆RAC and their sides are in proportion: *CA/MA* = *RC/PM*

(d) ∆PCM ~ ∆SCA and their sides are in proportion: *CM/CA* = *PM/AS*

Multiply each side of (a), (b), (c), and (d) together:

(AN/NB)(BL/LC)(CA/MA)(CM/CA) = *(AS/BP)(BP/CR)(RC/PM)(PM/AS)*

Cancel like terms in the numerators and denominators to prove Ceva's Theorem:

(AN/NB)(BL/LC)(C̶A̶/MA)(CM/C̶A̶) = *(A̶S̶/B̶P̶)(B̶P̶/C̶R̶)(R̶C̶/P̶M̶)(P̶M̶/A̶S̶)*

(AN/NB)(BL/LC)(CM/MA) = 1

Principles used in this informal proof include the following:

AA similarity postulate: If two angles of one triangle are congruent to two angles of another triangle, the triangles are similar.

Parallel lines cut by a transversal have congruent corresponding angles and congruent alternate interior angles.

Corresponding sides of similar triangles have the same proportion.

Triangle proportionality theorem (also called the side-splitter theorem): If a line is parallel to one side of a triangle and intersects the other two sides, it divides the sides it intersects proportionally.

• *Cevians* of a triangle that are *concurrent* (meet at one point) include medians, altitudes, and interior angle bisectors. Ceva's Theorem can be used to prove that the three medians, altitudes, and interior angle bisectors of a triangle are concurrent.

• **Example:** You are designing a lavish hotel. The foundation of the hotel is a triangle. There will be three grand hallways on the main floor beginning at the midpoint of each side that cross each other and continue to each of the three corners or vertexes of the triangle. In ∆ABC (below), the hallways are *AL*, *BM*, and *CN*. You need the three hallways to meet precisely at one point (P) where you plan to construct a magnificent fountain surrounded by the main lobby. Before submitting your plans, you must prove that the three hallways will intersect at the fountain. To do this, you realize that the hallways are *medians of a triangle*; therefore, you must prove that the ***medians of this or any triangle are concurrent*** (meet at a single point). In the figure of △ABC, *AL*, *BM*, and *CN* are the medians (hallways).

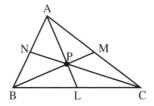

We can use ***Ceva's Theorem to prove the medians concurrent theorem***, which states that *the three medians of a triangle are concurrent.*

Because *AL*, *BM*, and *CN* are the medians of △ABC, then

$AN = NB$, $BL = LC$, and $CM = MA$

Multiply these three equations:

$(AN)(BL)(CM) = (NB)(LC)(MA)$

Divide both sides by $(NB)(LC)(MA)$:

$$\frac{(AN)(BL)(CM)}{(NB)(LC)(MA)} = 1$$

Because $AN = NB$, $BL = LC$, and $CM = MA$ each cancel, then:

$(AN/NB)(BL/LC)(CM/MA) = 1$

By Ceva's Theorem, if this equation is true, then the cevians, or in this case medians, *AL*, *BM*, and *CN*, are concurrent and the hallways meet at one point.

• **Example:** Prove the ***angle bisectors concurrent theorem***: *Interior angle bisectors of any triangle are concurrent.* In the figure of △ABC, *AL*, *BM*, and *CN* are the interior angle bisectors.

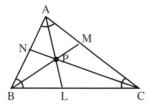

We can use *Ceva's Theorem* and the *triangle angle bisector theorem* to prove that the interior angle bisectors of any triangle are concurrent. The *triangle angle bisector theorem* states that a bisector of an angle of a triangle divides the opposite side (of the triangle) into segments, which are proportional to the adjacent sides.

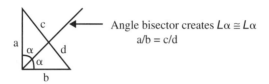

Angle bisector creates $L\alpha \cong L\alpha$
a/b = c/d

We are given *AL*, *BM*, and *CN* are the interior angle bisectors of △ABC, and we can use the *triangle angle bisector theorem* for each angle:

$AN/NB = AC/BC$, $BL/LC = AB/AC$, and $CM/MA = BC/AB$

Multiply these three equations:

$(AN/NB)(BL/LC)(CM/MA) = (AB/AC)(AC/BC)(BC/AB)$

Cancel like terms in numerators and denominators:

$(AN/NB)(BL/LC)(CM/MA) = (\cancel{AB}/\cancel{AC})(\cancel{AC}/\cancel{BC})(\cancel{BC}/\cancel{AB})$
$(AN/NB)(BL/LC)(CM/MA) = 1$

By Ceva's Theorem, if this equation is true, then the cevians, or, in this case, interior angle bisectors, *AL*, *BM*, and *CN*, are concurrent.

Note that this can be proved without using Ceva's Theorem, but the proof, which you can find in many a geometry books, is much more laborious.

• **Example:** Prove the *altitudes concurrent theorem*: Altitudes of any triangle are concurrent. In △ABC, *AL*, *BM*, and *CN* are the altitudes.

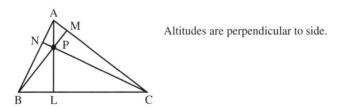

Altitudes are perpendicular to side.

We can use *Ceva's Theorem* and properties of similar triangles (proportional sides) to prove that the *altitudes* of any triangle are concurrent. First we use *AA similarity triangle postulate* to demonstrate which triangles are similar, and then write the proportions of each triangle pair. The

AA similarity triangle postulate states that if *two* angles of one triangle are congruent to *two* angles of another triangle, then the triangles are *similar.* (The symbol ~ denotes that triangles are similar. See Section 4.6.)

∠BAM ≅ ∠NAC (same angle) and ∠AMB ≅ ∠ANC (right angles since alts ⊥). Using *AA similarity triangle postulate*, ΔANC ~ ΔAMB; therefore, sides are in proportion: *AN/MA = AC/AB*

∠ABL ≅ ∠NBC (same angle) and ∠BLA ≅ ∠BNC (right angles since alts ⊥). Using *AA similarity triangle postulate*, ΔBLA ~ ΔBNC; therefore, sides are in proportion: *BL/NB = AB/BC*

∠MCB ≅ ∠ACL (same angle) and ∠CMB ≅ ∠CLA (right angles since alts ⊥). Using *AA similarity triangle postulate*, ΔCMB ~ ΔCLA; therefore, sides are in proportion: *CM/LC = BC/AC*

Multiply these three equations:

$(AN/MA)(BL/NB)(CM/LC) = (AC/AB)(AB/BC)(BC/AC)$

Cancel like terms:

$(AN/MA)(BL/NB)(CM/LC) = (\cancel{AC}/\cancel{AB})(\cancel{AB}/\cancel{BC})(\cancel{BC}/\cancel{AC})$

$(AN/MA)(BL/NB)(CM/LC) = 1$

By Ceva's Theorem, if this equation is true, then the cevians, or in this case altitudes, *AL*, *BM*, and *CN*, are concurrent.

Note: In this problem, we are using the fact that altitude lines are perpendicular to their base and form right angles.

4.4 Inequalities and Triangles

Properties of inequalities that are extensions of axioms

(Axioms are described in Section 1.4. *Key Axioms and Postulates*. Also, see *Master Math: Basic Math and Pre-Algebra*, Section 1.19, for an introduction to inequalities.)

• *Transitive inequality property*:

If x > y and y > z, then x > z
If x < y and y < z, then x < z
If 3 > 2 and 2 > 1, then 3 > 1

• *Substitution axiom also applies to inequalities*: A quantity may be substituted for an equal quantity in any expression or equation.

If x = y, then x can be substituted for y in any inequality.

- *Partition inequality property*: The whole is greater than any of its individual parts.

If $x = y + z$, then $x > y$ and $x > z$ providing $y > 0$ and $z > 0$.

If $AB = 3$, $BC = 2$, $CD = 2$, and $AD = 7$
then, $AD > AB$, $AD > BC$, and $AD > CD$.

A   B C D
   3   2 2

- *Addition inequality property*:

If $x > y$, then $(x + z) > (y + z)$
If $x < y$, then $(x + z) < (y + z)$
If $x > y$ and $z > q$, then $(x + z) > (y + q)$
If $x < y$ and $z < q$, then $(x + z) < (y + q)$
If $3 > 2$ and $z = 5$, then $(3 + 5) > (2 + 5)$

- *Subtraction inequality property*:

If $x > y$, then $(x - z) > (y - z)$
If $x < y$, then $(x - z) < (y - z)$
If $3 > 2$ and $z = 1$, then $(3 - 1) > (2 - 1)$

- *Multiplication inequality property*:

If $x > y$ and $z > 0$, then $(x)(z) > (y)(z)$
If $x < y$ and $z > 0$, then $(x)(z) < (y)(z)$
If $3 > 2$ and $z = 4$, then $(3)(4) > (2)(4)$

- *Division inequality property*:

If $x > y$ and $z > 0$, then $(x/z) > (y/z)$
If $x < y$ and $z > 0$, then $(x/z) < (y/z)$
If $4 > 3$ and $z = 2$, then $(4/2) > (3/2)$

- **Example:** In $\triangle ABC$, D is the midpoint of AB, E is the midpoint of AC, and $AD < AE$. Prove that $AB < AC$.

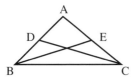

Given: $\triangle ABC$ with D as midpoint of AB, E as midpoint of AC, and $AD < AE$.

Prove: $AB < AC$.

Plan: By definition of midpoint, $AB = (2)AD$ and $AC = (2)AE$. By multiplication inequality property, $AD < AE$ becomes $(2)AD < (2)AE$. Then, by substitution $AB < AC$.

Proof:

Statements	Reasons
1. ΔABC with D as midpoint of *AB* and E as midpoint of *AC*.	1. Given.
2. *AB* = (2)*AD*; *AC* = (2)*AE*.	2. Definition of midpoint.
3. *AD* < *AE*.	3. Given.
4. (2)*AD* < (2)*AE*.	4. Multiplication inequality property (for #3): If $x < y$ and $z > 0$ then $(xz) < (yz)$.
5. *AB* < *AC*.	5. Substitution axiom: A quantity is substituted for an equal quantity. (for #2 & #4)

• ***Triangle inequality theorem***: The sum of the lengths of any two sides of a triangle is always greater than the length of the third side.

In other words, the length of one side of a triangle is always less than the sum of the lengths of the other two sides.

$$a + b > c, \ b + c > a, \ a + c > b$$

• **Example:** In a triangle, if one side length is 9 and a second side length is 14, which of the following could possibly be the length of the third side: (a) 3, (b) 6, or (c) 22?

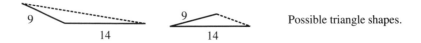

Possible triangle shapes.

Because the triangle inequality theorem tells us that the length of one side of a triangle is always less than the sum of the lengths of the other two sides, in this example, we can determine whether (a), (b), or (c) could possibly be the third side by calculating whether the sum of the lengths of the two shorter sides is greater than the length of the longer side for each case.

(a) 3, 9, and 14? Is 3 + 9 > 14? No 12 not> 14, so 3 *can't* be the third side.

(b) 6, 9, and 14? Is 6 + 9 > 14? Yes 15 > 14, so 6 *can* be the third side.

(c) 22, 9, and 14? Is 9 + 14 > 22? Yes 23 > 22, so 22 *can* be the third side.

Therefore, either 6 or 22 could be the third side.

We can check these results by finding ranges of the third side x by solving the possible inequality for x:

Solve $9 + 14 > x$: $23 > x$
Solve $9 + x > 14$: $x > 5$
Solve $14 + x > 9$: $x > -5$ (not relevant)

The length of the third side will be greater than 5 and less than 23.

• **In a triangle, the largest side is opposite the largest angle, the smallest side is opposite the smallest angle, and the middle-length side is opposite the middle-size angle.**

Large side is opposite large angle.
Medium side is opposite medium angle.
Small side is opposite small angle.

• *Unequal sides theorem*: If two *sides* of a triangle are *unequal*, then the measures of the angles opposite those sides are *unequal*, and the greater angle is opposite the longer side.

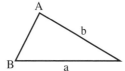

If side a > side b, then $\angle$A > $\angle$B.
If $\angle$A > $\angle$B, then side a > side b.

Unequal angles theorem: If two *angles* of a triangle are *unequal*, then the measures of the sides opposite those angles are *unequal*, and the longer side is opposite the greater angle.

• **Example:** Which side of the triangle is shortest?

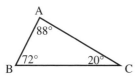

Side *AB* is opposite the smallest angle and therefore is the shortest side.

• **Example:** Prove *unequal sides theorem*, which states that if two *sides* of a triangle are *unequal*, then the measures of the angles opposite those sides are *unequal*, and the greater angle is opposite the longer side.

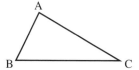
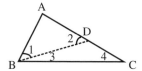

Given: ΔABC having $AC > AB$.

Prove: $\angle ABC > \angle ACB$.

Plan: Create isosceles triangle ΔABD and show that $\angle ABC > \angle 2$ and $\angle 2 > \angle 4$; therefore, $\angle ABC > \angle ACB$.

Proof:

Statements	Reasons
1. ΔABC having $AC > AB$.	1. Given.
2. Identify D so that $AB \cong AD$.	2. Ruler postulate.
3. Draw BD making isosceles ΔABD.	3. Between 2 points exists one line.
4. $\angle 1 \cong \angle 2$.	4. If 2 sides of a Δ are ≅, then angles opposite those sides are ≅.
5. $\angle ABC = \angle 1 + \angle 3$.	5. Angle addition postulate.
6. $\angle ABC > \angle 1$.	6. The whole is greater than one of its parts.
7. $\angle ABC > \angle 2$.	7. Substitution of #6 with #4.
8. $\angle 2 > \angle 4$.	8. $\angle 2$ is an exterior ∠ to ΔDBC, and is greater than either remote interior angle $\angle 3$ or $\angle 4$, so that $\angle 2 > \angle 4$ and $\angle 2 > \angle 3$.
9. $\angle ABC > \angle 4$, or equivalently $\angle ABC > \angle ACB$.	9. Transitive prop of inequality: If $\angle a > \angle b$ and $\angle b > \angle c$, then $\angle a > \angle c$, or in this case, $\angle ABC > \angle 2$ and $\angle 2 > \angle 4$, so $\angle ABC > \angle 4$.

• **SAS inequality theorem**: If *two sides* of one triangle are congruent to the corresponding two sides of another triangle, but the included angle of the first triangle is greater than the included angle of the second triangle, then the third side of the first triangle is longer than the third side of the second triangle. (This is a modification of the SAS congruence postulate described in Section 4.5. *Congruent Triangles*.)

If $AB \cong DE$, $AC \cong DF$, and $\angle A > \angle D$, then $BC > EF$

• *SSS inequality theorem*: If *two sides* of one triangle are congruent to the corresponding two sides of another triangle, but the third side of the first triangle is longer than the third side of the second triangle, then the included angle of the first triangle is greater than the included angle of the second triangle. (This is a modification of the SSS congruence postulate described in Section 4.5. *Congruent Triangles*.)

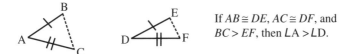

If $AB \cong DE$, $AC \cong DF$, and $BC > EF$, then $LA > LD$.

• *Exterior angle inequality theorem*: The measure of an exterior angle of a triangle is greater than the measure of either non-adjacent remote interior angle.

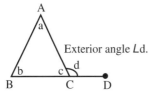

Exterior angle Ld.

$Ld > La$ and $Ld > Lb$.
La and Lb are non-adjacent remote interior angles to Ld.

Remember, an *exterior angle of a triangle* is an angle formed between the extension of one of the sides of the triangle and the outside of the triangle. Also remember $La + Lb = Ld$ from *exterior/remote interior angle theorem*. (These two theorems are introduced on page 97.)

• In a right triangle, because the sum of all the angles is 180° and the right angle is 90°, each acute angle will always be smaller than 90°. Therefore, the side opposite the 90° angle, which is the hypotenuse, will always be the longest side.

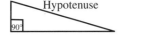

The hypotenuse is the longest side of a right triangle (opposite the largest angle).

4.5 Congruent Triangles

• *Congruent triangles* **have the same size and same shape so that the** *three corresponding sides are of equal length* **and the** *three corresponding angles have equal measure*. **Congruent triangles are exact duplicates of each other. The** *symbol for congruent* **is** $\cong$.

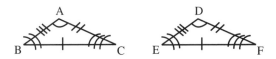

(Congruence indicated with tick marks and arcs.)

If $\triangle ABC \cong \triangle DEF$, then:

Corresponding sides are congruent: $AB \cong DE$; $BC \cong EF$; $AC \cong DF$.
Corresponding angles are congruent: $\angle A \cong \angle D$; $\angle B \cong \angle E$; $\angle C \cong \angle F$.

Because congruent triangles have the same *size*, corresponding *sides* are congruent.

Because congruent triangles have the same *shape*, corresponding *angles* are congruent.

• *Corresponding parts of congruent triangles have the same measure and are therefore congruent to each other*. *Corresponding parts* include sides, angles, altitudes, medians, etc.

• *Congruent corresponding parts triangle theorem*: **If two triangles are congruent, then their corresponding parts are congruent.**

If $\triangle ABC \cong \triangle DEF$, then the corresponding parts are:

$\angle A \cong \angle D$, $\angle B \cong \angle E$, and $\angle C \cong \angle F$.
$AB \cong DE$, $BC \cong EF$, and $AC \cong DF$.

• Corresponding sides lie opposite corresponding angles, and corresponding angles lie opposite corresponding sides.

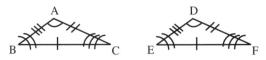

If $AB \cong DE$, then $\angle C \cong \angle F$, or if $\angle C \cong \angle F$, then $AB \cong DE$.
If $BC \cong EF$, then $\angle A \cong \angle D$, or if $\angle A \cong \angle D$, then $BC \cong EF$.
If $AC \cong DF$, then $\angle B \cong \angle E$, or if $\angle B \cong \angle E$, then $AC \cong DF$.

• When you are describing congruent triangles, *list the corresponding vertex angles in the same order*. Corresponding vertices are the vertices that overlap if the two triangles are placed with one over the other, matching the sides and vertices. For example, in the triangles below, if ∠A corresponds to ∠D, ∠B corresponds to ∠E, and ∠C corresponds to ∠F, we should write: ΔABC ≅ ΔDEF (instead of ΔABC ≅ ΔFDE).

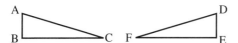

How Do We Know If Two Triangles Are Congruent?

• There are certain postulates and theorems that can be used to determine whether two triangles are congruent. *To determine congruence for two triangles, we need to identify at least three pairs of congruent parts which must include at least one pair of known congruent side lengths.* The three corresponding parts which must be congruent may be: three sides (SSS); two sides and their included angle (SAS); two angles and their included side (ASA); or two angles and a non-included side (AAS). If the triangles are *right triangles*, the right angles are automatically congruent, so there are only two other pairs of corresponding parts that need to be congruent for the right triangles to be congruent. These include the hypotenuse and one leg (HL); the hypotenuse and one acute angle (HA); one leg and one acute angle (LA); or the two legs (LL). These congruence requirements are explained in the following paragraphs.

Note: In the following drawings **tick marks** and **arcs** indicate congruent sides and angles.

• *SSS congruence postulate*: (Side-Side-Side Postulate) **If the *three sides* of one triangle are congruent to the corresponding three sides of another triangle, then the triangles are congruent.**

Three corresponding sides are congruent. Therefore the triangles are congruent.

• *SAS congruence postulate*: (Side-Angle-Side Postulate) **If *two sides and the included angle* of one triangle are congruent to the corresponding two sides and the included angle of another triangle, then the triangles are congruent.**

Two corresponding sides and the included corresponding angles are congruent. Therefore the triangles are congruent.

• *ASA congruence postulate*: (Angle-Side-Angle Postulate) **If *two angles and the included side* of one triangle are congruent to the corresponding two angles and the included side of another triangle, then the triangles are congruent.**

Two corresponding angles and the included corresponding sides are congruent. Therefore, the triangles are congruent.

• *AAS (SAA) congruence theorem*: (Angle-Angle-Side Theorem) **If *two angles and a not-included side* of one triangle are congruent to the corresponding two angles and a not-included side of another triangle, then the triangles are congruent.**

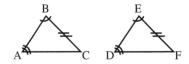

Two corresponding angles and not-included corresponding sides are congruent. Therefore, the triangles are congruent.

Informal proof of AAS congruence theorem:

Given: △ABC and △DEF; ∠A ≅ ∠D; ∠B ≅ ∠E; and $BC ≅ EF$.

Prove: △ABC ≅ △DEF.

Proof: First, because ∠A ≅ ∠D and ∠B ≅ ∠E, then ∠C ≅ ∠F. This is true because if two angles in one triangle are congruent to two angles in a second triangle, the third angles must be congruent because the sum of the angles in each triangle is 180°.

Second, because $BC ≅ EF$ and they are the included sides between congruent angles ∠B and ∠E and congruent angles ∠C and ∠F, then using the ASA congruence postulate above, △ABC ≅ △DEF.

• *HL right triangle congruence theorem*: (Hypotenuse-Leg Theorem) If the *hypotenuse and one leg* of one right triangle are congruent to the hypotenuse and one leg of another right triangle, then the right triangles are congruent.

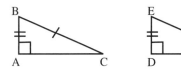

One pair of corresponding legs, the right angles, and the hypotenuses are congruent. Therefore, the triangles are congruent.

Proof of *HL right triangle congruence theorem*:

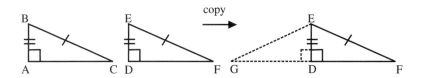

Given: ΔABC and ΔDEF are right triangles; ∠A and ∠D are right angles; *AB* ≅ *DE*; *BC* ≅ *EF*.

Prove: ΔABC ≅ ΔDEF.

Proof:

Statements	Reasons
1. Extend line through *DF*.	1. Two points determine one line.
2. Mark G on *DF* so that *GD* ≅ *AC*.	2. Ruler postulate.
3. Draw line between E & G.	3. Through two points exists one line.
4. ∠A & ∠EDG are right angles.	4. Right triangles have a right angle.
5. ∠A ≅ ∠EDG.	5. Right angles are congruent.
6. *AB* ≅ *DE*.	6. Given.
7. ΔABC ≅ ΔDEG.	7. SAS ≅ post: If 2 sides (#2, #6) and included ∠ (#5) of one Δ are ≅ to corresponding parts of other Δ, Δs are ≅.
8. *BC* ≅ *EG*.	8. Corresponding parts of ≅ Δs are ≅.
9. *BC* ≅ *EF*.	9. Given.
10. *EG* ≅ *EF*.	10. Transitive: Substitute #8 and #9.
11. ∠G ≅ ∠F.	11. In ΔGEF: if 2 sides (*EG* ≅ *EF*) are ≅, ∠s opposite ≅ sides are ≅. (Isosceles Δ)
12. ∠G ≅ ∠C.	12. Corresponding parts of ≅ Δs are ≅. (#7)
13. ∠F ≅ ∠C.	13. Transitive: Substitute #11 and #12.
14. ∠A & ∠D are right angles.	14. Given.
15. ∠A ≅ ∠D.	15. Right angles are congruent.
16. ΔABC ≅ ΔDEF.	16. AAS congruence thm: If 2 ∠s (#13, #15) and a not-included side (#6) of one Δ are ≅ to corresponding parts of other Δ, Δs are ≅.

• *Note*: *Having three congruent angles (AAA) does not guarantee that two triangles are congruent (just similar). Also, having congruent side-side-and not-included-angle (SSA) does not guarantee that two triangles are congruent.*

• Following are additional theorems specific for *right triangles* that are similar to above described AAS theorem and ASA and SAS postulates for any type of triangle.

• **HA right triangle congruence theorem**: (Hypotenuse-Angle Theorem) If the *hypotenuse and one acute angle* of one right triangle are congruent to the hypotenuse and one acute angle of another right triangle, then the right triangles are congruent.

One pair of corresponding acute angles, the right angles, and the hypotenuses are congruent. Therefore the triangles are congruent. (This is similar to AAS.)

• **LA right triangle congruence theorem**: (Leg-Angle Theorem) If *one leg and one acute angle* of one right triangle are congruent to one leg and one acute angle of another right triangle, then the right triangles are congruent.

One pair of corresponding acute angles, the right angles, and a pair of corresponding legs are congruent. Therefore the triangles are congruent. (This is similar to AAS or ASA.)

• **LL right triangle congruence theorem**: (Leg-Leg Theorem) If the *two legs* of one right triangle are congruent to the two legs of another right triangle, then the right triangles are congruent.

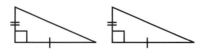

Both pairs of corresponding legs and the right angles are congruent. Therefore the triangles are congruent. (This is similar to SAS.)

• *Note*: **HL** and **LL** can be summarized as: If two corresponding sides of two *right triangles* are congruent, the third corresponding sides are also congruent, and the triangles are ***congruent***.

• **Example:** If $AC \cong BD$, $AB \perp BC$, and $BC \perp CD$, then are triangles $\triangle ABC$ and $\triangle DCB$ congruent?

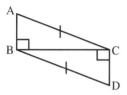

$\triangle ABC$ and $\triangle DCB$ are right triangles because perpendicular lines form right angles. These two right triangles ($\triangle ABC$ and $\triangle DCB$) have congruent right angles ($\angle ABC \cong \angle DCB$), congruent hypotenuses ($AC \cong BD$), and a common and congruent leg ($BC \cong BC$). Therefore, triangles $\triangle ABC$ and $\triangle DCB$ are congruent because of the *HL right triangle congruence theorem*, which states that if the *hypotenuse and one leg* of one right triangle are congruent to the hypotenuse and one leg of another right triangle, then the right triangles are congruent.

• **Example:** You are a molecular biologist researching local plants for their potential usefulness as life-saving drugs. You are in a remote area of South America with a few of your research colleagues. A nearby thought-to-be extinct volcano begins to erupt. As you are escaping the lava flows, a chasm opens up in the ground between you an your escape route. There is an old burned down shed nearby with a long undamaged plank of wood lying next to it. One of your colleagues suggests pushing the plank up on its end and across the chasm to use as a bridge. However, because there is only one plank, you don't want to lose it over the edge at this location if it is too wide and the plank is too short. You may have to drag it someplace else where the chasm is narrower. You need to quickly determine the distance across the chasm at your location. How will you measure the distance across the chasm?

Suddenly, you have a brilliant idea from your geometry class. You will use congruent triangles!

1. Select a <u>T</u>ree on the edge of the far side of the chasm at the narrowest point you can see and mark your side (across from the tree) with one of your colleagues <u>A</u>lan.

2. Use the plank to measure one plank-length at a right angle from the tree along your side of the chasm. Have <u>B</u>eth stand there.

3. Use the plank to measure one more plank-length further in the same direction along your side of the chasm. Have C̲alvin stand there.

4. Then turn with your back to the chasm and walk away from the chasm until you can see that Beth is exactly in between you and the Tree on the other side of the chasm. Have D̲awn stand there.

5. The distance between C̲alvin and D̲awn will be equal to the distance between A̲lan and the T̲ree across the chasm (because you have formed two congruent triangles ΔABT and ΔCBD).

6. Hold the plank between C̲alvin and D̲awn to determine if it is long enough to reach across the chasm (A̲lan to T̲ree)... It is!

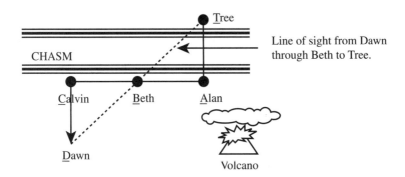

You have saved the day! Your fellow grateful adventurers want you to show them how this idea using your congruent triangles worked.

Once you are back safely at camp, you explained that what you created was two congruent triangles, ΔABT ≅ ΔCBD. These triangles are congruent because ∠A and ∠C are both right angles, the plank lengths CB and AB are of equal length, and ∠ABT ≅ ∠CBD because they are vertical angles. By the *ASA (angle-side-angle) congruence postulate*, we know that the triangles are congruent and therefore have equivalent corresponding sides and angles, which means side CD is equal to side AT. Then, because the plank was longer than CD, you knew it would reach across AT.

Then, you show your formal proof:

Given: ∠ABT and ∠CBD are vertical angles; $CB ≅ AB$; $CD ⊥ CB$; $AT ⊥ AB$.

Prove: $CD ≅ AT$ and plank will reach across chasm AT.

Plan: Show ΔABT ≅ ΔCBD using ASA congruence postulate. Because corresponding parts of congruent triangles are congruent, then $CD ≅ AT$.

Proof:

Statements	Reasons
1. $\angle ABT$ and $\angle CBD$ are vertical $\angle$s.	1. Given by marking off your side.
2. $\angle ABT \cong \angle CBD$.	2. Vertical angles are congruent.
3. $CB \cong AB$.	3. Given by measuring plank.
4. $CD \perp CB$; $AT \perp AB$.	4. Given by turning 90°.
5. $\angle A$ and $\angle C$ are right angles.	5. Perpendicular lines form right angles.
6. $\angle A \cong \angle C$.	6. Right angles are congruent.
7. $\triangle ABT \cong \triangle CBD$.	7. ASA congruence postulate: If 2 $\angle$s (#2, #6) and included side (#3) of one $\triangle$ are $\cong$ to corresponding parts of other $\triangle$, $\triangle$s $\cong$.
8. $CD \cong AT$.	8. Corresponding parts of $\cong$ $\triangle$s are $\cong$.
9. Plank will reach across chasm AT.	9. Because plank > CD, then plank > AT.

• **Example:** You have a large quadrilateral-shaped auditorium in which you are planning to install a removable wall across the diagonal (as depicted in the figure). It is important that the two pairs of angles formed by the wall (diagonal) have equal measures. Assuming that both pairs of opposite sides of the quadrilateral auditorium are congruent, prove that congruent angles are formed by the diagonal wall.

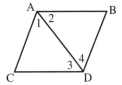

Given: Quadrilateral ABCD; $AB \cong CD$; $AC \cong BD$; AD is the diagonal.

Prove: $m\angle 1 = m\angle 4$ and $m\angle 2 = m\angle 3$.

Plan: Show that $\triangle ADC \cong \triangle DAB$ using the SSS congruence postulate. Then, $m\angle 1 = m\angle 4$ and $m\angle 2 = m\angle 3$ because corresponding angles of congruent triangles are congruent.

Proof:

Statements	Reasons
1. $AB \cong CD$; $AC \cong BD$.	1. Given.
2. $AD \cong AD$.	2. Identity or reflexive axiom: Any quantity is equal to itself.
3. $\triangle ADC \cong \triangle DAB$.	3. SSS congruence postulate: If the 3 sides of one $\triangle$ are $\cong$ to the 3 sides of another $\triangle$, then the $\triangle$s are $\cong$.
4. $\angle 1 \cong \angle 4$ and $\angle 3 \cong \angle 2$.	4. Corresponding angles of $\cong$ $\triangle$s are $\cong$.

• *Note*: The *parallelogram sides theorem* states that a quadrilateral is a parallelogram if both pairs of opposite *sides* are congruent. (See Chapter 5, Section 5.)

• When working through a proof, in order to show congruence between sides or angles of two triangles, it is often useful to first show that the triangles themselves are congruent. Then it follows that the parts in question (side lengths or angle measurements) are also congruent because *corresponding parts of congruent triangles are congruent.*

• **Example:** If the diagonals of a quadrilateral bisect each other, prove that both pairs of opposite sides are congruent.

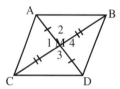

Given: Quadrilateral ABCD; Diagonal *AD bisects BC* at M; Diagonal *BC bisects AD* at M.

Prove: $AB \cong CD$ and $AC \cong BD$.

Plan: Show that $\triangle AMC \cong \triangle DMB$ and $\triangle AMB \cong \triangle DMC$ using the SAS congruence postulate. Then, $AB \cong CD$ and $AC \cong BD$ because corresponding parts of congruent triangles are congruent.

Proof:

Statements	Reasons
1. In ABCD diagonals bisect.	1. Given.
2. $AM \cong MD$; $CM \cong MB$.	2. Def. bisector: Divides into two $\cong$ parts.
3. $L1 \cong L4$ and $L2 \cong L3$.	3. Def: Vertical angles are congruent.
4. $\triangle AMC \cong \triangle DMB$; $\triangle AMB \cong \triangle DMC$.	4. SAS congruence postulate: If 2 sides and included angle of one $\triangle$ are $\cong$ to corresponding parts of other $\triangle$, $\triangle$s are $\cong$.
5. $AB \cong CD$ and $AC \cong BD$.	5. Corresponding parts of $\cong$ $\triangle$s are $\cong$.

• **Example:** Prove that $CB \cong CD$, given $EB \cong ED$ and $AB \cong AD$.

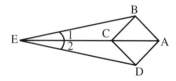

Given: $EB \cong ED$; $AB \cong AD$.

Prove: $CB \cong CD$.

Plan: First show $\triangle EBA \cong \triangle EDA$ using SSS postulate, resulting in $\angle 1 \cong \angle 2$ because corresponding parts of congruent triangles are congruent.

Then show $\triangle EBC \cong \triangle EDC$ using SAS postulate, resulting in $CB \cong CD$ because corresponding parts of congruent triangles are congruent.

Proof:

Statements	Reasons
1. $EB \cong ED$; $AB \cong AD$.	1. Given.
2. $EA \cong EA$.	2. Identity or reflexive axiom: Self = self.
3. $\triangle EBA \cong \triangle EDA$.	3. SSS congruence postulate: If 3 sides of one $\triangle$ are $\cong$ to corresponding parts of other $\triangle$, $\triangle$s are $\cong$. (#1 and #2)
4. $\angle 1 \cong \angle 2$.	4. Corresponding parts of $\cong$ $\triangle$s are $\cong$.
5. $EC \cong EC$.	5. Identity or reflexive axiom: Self = self.
6. $\triangle EBC \cong \triangle EDC$.	6. SAS congruence post: If 2 sides and included $\angle$ of one $\triangle$ are $\cong$ to corresponding parts of other $\triangle$, $\triangle$s $\cong$. (#1 $EB \cong ED$, #4, #5.)
7. $CB \cong CD$.	7. Corresponding parts of $\cong$ $\triangle$s are $\cong$.

• **Example:** Given AD and BC bisect each other at point E, prove that $AB \parallel CD$.

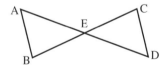

Given: AD and BC bisect each other at point E.

Prove: $AB \parallel CD$.

Plan: Show $\triangle AEB \cong \triangle DEC$ using SAS congruence postulate (side-angle-side). Because corresponding angles of congruent triangles are congruent, $\angle A \cong \angle D$ and $\angle B \cong \angle C$. $AB \parallel CD$ by parallel transversal postulate, because two lines are parallel if, when cut by a transversal (AD or BC), a pair of corresponding angles ($\angle A$ & $\angle D$ or $\angle B$ & $\angle C$) is congruent.

Proof:

Statements	Reasons
1. Segments *AD* and *BC* bisect each other at point E.	1. Given.
2. *AE* ≅ *ED* and *BE* ≅ *EC*.	2. Def: Bisector divides into ≅ parts.
3. ∠AEB ≅ ∠DEC.	3. Vertical angles are ≅ to each other.
4. ΔAEB ≅ ΔDEC.	4. SAS congruence post: If two ∠s and included side of one Δ are ≅ to corresponding parts of other Δ, Δs are ≅. (#3 and #2)
5. ∠A ≅ ∠D and ∠B ≅ ∠C.	5. Corresponding angles of ≅ Δs are ≅.
6. Segment *AB* ∥ segment *CD*.	6. Two lines are parallel if, when intersected by a transversal (*AD* or *BC*), a pair of corresponding angles (∠A ≅ ∠D or ∠B ≅ ∠C) are ≅.

• **Example:** In the drawing prove that *AB* ≅ *CD*, given E is midpoint of *AD* and ∠A ≅ ∠D and ∠1 and ∠2 are vertical angles.

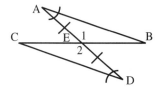

Given: ∠A ≅ ∠D; ∠1 and ∠2 are vertical angles; E is midpoint of *AD*.

Prove: *AB* ≅ *CD*.

Plan: Show ΔABE ≅ ΔDCE by ASA (angle-side-angle) congruence postulate using ∠1 ≅ ∠2, ∠A ≅ ∠D, and *AE* ≅ *ED*. Then, because corresponding parts of congruent triangles are congruent, it follows that *AB* ≅ *CD*.

Proof:

Statements	Reasons
1. ∠1 and ∠2 are vertical angles.	1. Given.
2. ∠1 ≅ ∠2.	2. Vertical angles are congruent.
3. E is midpoint of *AD*.	3. Given.
4. *AE* ≅ *ED*.	4. The midpoint of a segment divides it into two congruent parts.
5. ∠A ≅ ∠D.	5. Given.
6. ΔABE ≅ ΔDCE.	6. ASA congruence post: If 2 ∠s and included side of one Δ are ≅ to corresponding parts of other Δ, Δs are ≅. (#2, #4, #5)
7. *AB* ≅ *CD*.	7. Corresponding parts of ≅ Δs are ≅.

• **Example:** Prove that $\triangle ABC \cong \triangle ADE$ given that $AB \cong AD$ and $AC \cong AE$.

Given: $AB \cong AD$ and $AC \cong AE$.

Prove: $\triangle ABC \cong \triangle ADE$.

Plan: We can prove congruence between these triangles using the SAS congruence postulate because we are given two congruent corresponding sides and the included angle A, which is a common angle of both triangles.

Proof:

Statements	Reasons
1. $AB \cong AD$.	1. Given.
2. $AC \cong AE$.	2. Given.
3. $LA \cong LA$.	3. Identity or reflexive axiom: Self = self.
4. $\triangle ABC \cong \triangle ADE$.	4. SAS congruence postulate: If 2 sides and included L of one $\triangle$ are $\cong$ to corresponding parts of other $\triangle$, $\triangle$s are $\cong$.

4.6 Similar Triangles: Congruent Angles and Sides in Proportion

• Geometric similarity is used by cartographers and architects to produce scaled drawings in order to create maps and design structures. For example, one inch on a scale drawing may represent 20 feet of a building or 20 miles on a map.

• Similar polygons, including *triangles*, have their corresponding *angles congruent* and their corresponding *sides in proportion*. Similar polygons, including triangles, have the same shape, but not necessarily the same size.

• *Definition*: **Two polygons are similar if and only if their vertices correspond so that the corresponding sides are in proportion (side ratios are equal) and the corresponding angles are congruent.**

Definitions and Properties of Similar Triangles

• The symbol for similar is: ~

For example, in the figure of similar triangles: $\Delta A_1B_1C_1 \sim \Delta A_2B_2C_2$

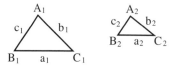

• *Definition*: *Corresponding angles of similar triangles are congruent.*

• *Definition*: *Corresponding sides of similar triangles are in proportion.* (Each pair of corresponding side lengths have equal ratios.)

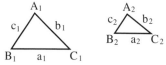

In the figure, $\angle A_1 \cong \angle A_2$, $\angle B_1 \cong \angle B_2$, $\angle C_1 \cong \angle C_2$, and corresponding sides are in proportion and therefore have equal ratios: $\dfrac{a_2}{a_1} = \dfrac{b_2}{b_1} = \dfrac{c_2}{c_1}$.

The ratios of corresponding sides of similar triangles are equal.

• *Note*: When you are describing similar triangles, list the corresponding angle vertexes in the same order. For example, if $\angle A$ corresponds to $\angle D$, $\angle B$ corresponds to $\angle E$, and $\angle C$ corresponds to $\angle F$, we should write: $\Delta ABC \sim \Delta DEF$.

• **Example:** In the following triangles, if $\angle A_1 \cong \angle A_2$ and $\angle B_1 \cong \angle B_2$, find c_1 and b_1.

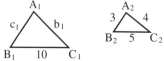

The angle-angle similarity postulate (listed following this example) states that if *two* angles of one triangle are congruent to *two* angles of another triangle, then the triangles are *similar*. Using this postulate, we find that $\Delta A_1B_1C_1 \sim \Delta A_2B_2C_2$. Because corresponding sides of similar triangles are in proportion, we can determine c_1 and b_1.

In $\Delta A_1B_1C_1$, side c_1 corresponds to side-length 3 in $\Delta A_2B_2C_2$ and b_1 corresponds to 4, just as 10 corresponds to 5.

The proportion is written: $c_1/3 = b_1/4 = 10/5$

This represents that the ratios of corresponding sides are equal.

Solve for c_1 and b_1 separately using each unknown-containing ratio equal to known ratio 10/5:

$c_1/3 = 10/5$
$c_1/3 = 2$
$c_1 = (2)(3) = 6$
$b_1/4 = 10/5$
$b_1/4 = 2$
$b_1 = (2)(4) = 8$

Therefore, $c_1 = 6$ and $b_1 = 8$, and the proportion of sides is:

$6/3 = 8/4 = 10/5 = 2/1 = 2:1$.

• *AA similarity triangle postulate*: **If *two* angles of one triangle are congruent to *two* angles of another triangle, then the triangles are similar.**

 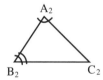

If $mLA_1 = mLA_2$ and $mLB_1 = mLB_2$, then $\Delta A_1B_1C_1 \sim \Delta A_2B_2C_2$

This is true because if two corresponding angles in each of the two triangles are equal to each other, then *the third angles will be equal* because the three angles in any planar triangle always sum to 180°.

• *Similarity/angle theorem*: **If all three pairs of *corresponding angles* in two triangles are *congruent* to each other, the two triangles are similar.**

$LA_1 \cong LA_2$
$LB_1 \cong LB_2$
$LC_1 \cong LC_2$

This follows from the *AA similarity triangle postulate* because if two angles of a triangle are known, the third angle can be determined because the sum of the three angles in a planar triangle is always $180°$.

• **SSS similarity theorem**: **If the *corresponding sides* of two triangles have the same *proportion* (but one triangle is larger than the other triangle), then the triangles are similar.**

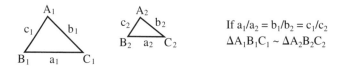

If $a_1/a_2 = b_1/b_2 = c_1/c_2$
$\triangle A_1 B_1 C_1 \sim \triangle A_2 B_2 C_2$

• **Example:** Using the two triangles in the above figure, if the sides are $c_1 = 15$, $b_1 = 21$, $a_1 = 27$, $c_2 = 5$, $b_2 = 7$, and $a_2 = 9$, determine whether the triangles are similar.

Write the proportions and determine whether the ratios of corresponding sides are equal:

$a_1/a_2 = b_1/b_2 = c_1/c_2$
$27/9 = 21/7 = 15/5$
$3 = 3 = 3$

Therefore, $\triangle A_1 B_1 C_1 \sim \triangle A_2 B_2 C_2$.

Remember, congruent triangles have both congruent corresponding angles and congruent corresponding sides.

• **SAS similarity theorem**: **If *one angle* of one triangle is congruent to an angle of another triangle and the corresponding *sides* that include these angles are in proportion, then the triangles are similar.**

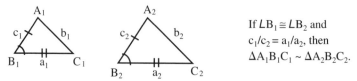

If $\angle B_1 \cong \angle B_2$ and $c_1/c_2 = a_1/a_2$, then $\triangle A_1 B_1 C_1 \sim \triangle A_2 B_2 C_2$.

Marked angles are congruent, marked sides are in *proportion*.

• Prove *SAS similarity theorem*.

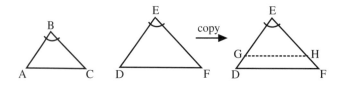

Given: △ABC and △DEF; $\angle$B $\cong$ $\angle$E; *AB/DE = BC/EF.*

Prove: △ABC ~ △DEF.

Plan: Draw *GE $\cong$ AB* and *GH $\parallel$ DF* forming △ABC $\cong$ △GEH. Then show △ABC ~ △DEF.

Proof:

Statements	Reasons
1. Select G so that *AB $\cong$ GE.*	1. Ruler postulate.
2. Draw *GH$\parallel$DF* through G forming △GEH inside △DEF.	2. Parallel line post: A line can be drawn $\parallel$ to a line through an external point.
3. *GE/DE = EH/EF.*	3. △ proportionality theorem: If a line is $\parallel$ to one side of a △, then it divides the other 2 sides proportionally.
4. *AB/DE = EH/EF.*	4. *AB $\cong$ GE* substitution from #1.
5. *AB/DE = BC/EF.*	5. Given.
6. *BC/EF = EH/EF.*	6. Transitive/Substitution of #4 & #5.
7. *BC = EH.*	7. Multiply both sides of #6 by *EF.*
8. $\angle$B $\cong$ $\angle$E.	8. Given.
9. △ABC $\cong$ △GEH.	9. SAS congruence post: If 2 sides and included $\angle$ of one △ is $\cong$ to corresponding parts of other △, △s are $\cong$. (#1, #8, #7)
10. $\angle$A $\cong$ $\angle$EGH.	10. Corresponding parts of $\cong$ △s are $\cong$.
11. $\angle$D $\cong$ $\angle$EGH.	11. Parallel lines (*GH$\parallel$DF*) cut by a transversal (*ED*) have $\cong$ corresponding $\angle$s.
12. $\angle$A $\cong$ $\angle$D.	12. Transitive axiom: values $\cong$ to the same are $\cong$ to each other. (#10 & #11)
13. △ABC ~ △DEF.	13. AA similarity post: If 2 $\angle$s of one △ are $\cong$ to 2 $\angle$s of other △, △s are similar. (#8, #12)

• **Example:** In the isosceles trapezoid, which pair of angles and what proportion can we use to show that △AEB ~ △CED?

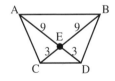

An isosceles trapezoid has two parallel sides and two sides having equal lengths (See Section 5.7. *Trapezoids*). To show similarity between △AEB and △CED we can use the *SAS similarity theorem* stating that if *one angle of one triangle is congruent to an angle of another triangle and the corresponding sides including these angles are in proportion, then the triangles are similar.* The pair of angles, $\angle$AEB and $\angle$CED, are congruent because they are vertical angles. The sides including these angles are: *EA $\cong$ EB* and *EC $\cong$ ED*, which are depicted as having equal lengths for each angle.

The sides of the triangles have the proportion: 9/3 = 9/3.

Therefore, because $\angle AEB \cong \angle CED$ and the sides including those angles of triangles $\triangle AEB$ and $\triangle CED$ are in proportion (9/3 = 9/3), then: $\triangle AEB \sim \triangle CED$.

• **Triangle to triangle similarity theorem**: Triangles that are similar to the same triangle are similar to each other.

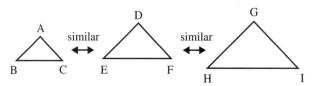

If $\triangle ABC \sim \triangle DEF$ and if $\triangle GHI \sim \triangle DEF$, then $\triangle ABC \sim \triangle GHI$.

• **Parallel sides similarity theorem**: Triangles are similar if their sides are respectively parallel to each other.

   If $AB \parallel DE$, $BC \parallel EF$, and $AC \parallel DF$,
                          then $\triangle ABC \sim \triangle DEF$.

• **Base angle isosceles theorem**: If one base angle of an *isosceles* triangle is congruent to a base angle of another isosceles triangle, then the triangles are similar.

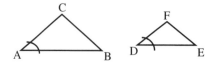

Prove *base angle isosceles theorem*:

Given: $\triangle ABC$ and $\triangle DEF$ are isosceles triangles; $\angle A \cong \angle D$.

Prove: $\triangle ABC \sim \triangle DEF$.

Plan: Because base angles of isosceles triangles are congruent, $\angle A \cong \angle B$ and $\angle D \cong \angle E$. Using the transitive axiom, $\angle B \cong \angle E$. Using the AA similarity triangle postulate and $\angle A \cong \angle D$ and $\angle B \cong \angle E$, then $\triangle ABC \sim \triangle DEF$.

Proof:

Statements	Reasons
1. $LA \cong LD$.	1. Given.
2. $\triangle ABC$ and $\triangle DEF$ are isosceles $\triangle$s.	2. Given.
3. $LA \cong LB$ and $LD \cong LE$.	3. Definition: Base angles of isosceles triangles are congruent.
4. $LB \cong LE$.	4. Transitive axiom: Angles congruent to the same angle, are $\cong$ to each other. (#1 & #3)
5. $\triangle ABC \sim \triangle DEF$.	5. AA similarity post: If 2 Ls of one $\triangle$ are $\cong$ to Ls of other $\triangle$, $\triangle$s are similar. (#1 & #4)

- **Example:** Determine the height of a tree using *similar triangles*.

We can find the height of a tree by setting a post of a known height next to the tree and measuring the length of the shadow. Then we can measure the length of the tree's shadow. The post and the tree and their shadows form similar triangles, as they are both right triangles and have equivalent right angles. They also have acute angles of the same measure at the shadow length where the sunlight strikes the ground. (We assume the Sun's rays are parallel as they strike the tree and the post.)

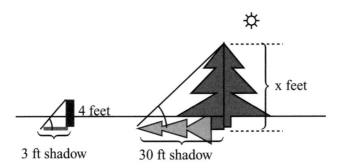

The height of the tree, x feet, is proportional to the height of the post, 4 feet. The length of the shadow of the tree, 30 feet, is proportional to the length of the shadow of the post, 3 feet. Therefore, the height of the tree represented by x can be determined using the proportion:

$$\frac{x}{4} = \frac{30}{3}, \text{ or } 3x = (30)(4), \text{ or } x = (30)(4)/3 = 40 \text{ feet}$$

Therefore, the height of the tree is 40 feet.

• **Example:** Determine the height of a telephone pole using a mirror and *similar triangles*.

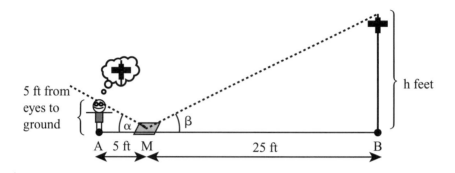

Determine the height of a telephone pole, h, using the following.

1. Place a mirror between you and the pole and adjust your position to view the top of the pole in the mirror:

   The distance between your eyes and the ground measures 5 feet.

2. Measure the distance between you and the mirror and the pole and the mirror:

   You and the mirror = *AM* = 5 feet.

   Pole and the mirror = *MB* = 25 feet.

3. Assume both you and the pole make a right angle with the ground, and therefore form congruent 90° angles: *m*LA = *m*LB = 90°

4. Angles α and β are congruent by optical reflection. If a beam of light was pointed at the mirror and bounced up to your eyes, the angle of incidence β and angle of reflection α would be the same: Lα ≅ Lβ

5. We have shown in #3 and #4 that there are two pairs of congruent angles, LA ≅ LB and Lα ≅ Lβ. Therefore, by the *angle-angle similarity postulate*, the triangle between you and the mirror and the triangle between the pole and the mirror are similar. In similar triangles the sides are in proportion so we can write that proportion to solve for the height of the pole:

   5/25 = 5/h, or 5h = (25)(5), or

   h = (25)(5) / (5) = 25

Therefore, the height of the telephone pole is 25 feet.

• **Example:** We know that corresponding angles of similar triangles are congruent. Using this and given that *AB* and *DE* are parallel explain why the two triangles are similar.

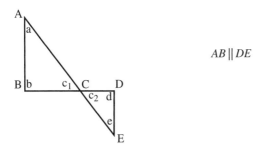

$AB \parallel DE$

Lc_1 and Lc_2 are vertical angles and therefore congruent.

La and Le are congruent because if parallel lines are cut by a transversal (AE), then the alternate interior angles are congruent.

Lb and Ld are congruent because if two corresponding angles of two triangles are congruent, then the third angles are also congruent. (They are also congruent because they are alternate interior angles of a transversal (BD) of parallel lines.)

Therefore, the two triangles are similar because all three corresponding angles are congruent. (*Similarity/angle theorem*: If all three pairs of corresponding angles in two triangles are congruent to each other, the two triangles are similar.)

• **Example:** Given that AB is parallel to DE and segments AE and BD intersect at point C, prove that $\triangle ABC$ is similar to $\triangle EDC$. Also, if DC is 10, BC is 15, and CE is 12, find AC.

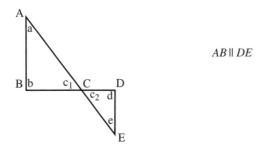

$AB \parallel DE$

Given: $AB \parallel DE$; segments AE and BD intersect at C.

Prove: $\triangle ABC \sim \triangle EDC$.

Plan: Show that because, $Lc_1 \cong Lc_2$ (vertical angles) and $La \cong Le$ (alternate interior angles of parallel lines cut by transversal), then $\triangle ABC \sim \triangle EDC$ (AA similarity triangle postulate).

Proof:

Statements	Reasons
1. *AE* and *BD* intersect at C.	1. Given.
2. $\angle c_1$ and $\angle c_2$ are vertical angles.	2. Definition of vertical angles.
3. $\angle c_1 \cong \angle c_2$.	3. Vertical angles are congruent.
4. $AB \parallel D\breve{E}$.	4. Given.
5. $\angle a \cong \angle e$.	5. Parallel lines cut by transversal (*AE*) have $\cong$ alternate interior $\angle$s.
6. $\triangle ABC \sim \triangle EDC$.	6. AA similarity $\triangle$ post: If 2 $\angle$s of one $\triangle$ are $\cong$ to 2 $\angle$s of other $\triangle$, $\triangle$s are similar. (#3, #5)

To determine *AC* if *DC* is 10, *BC* is 15, and *CE* is 12, we can use the fact that in similar triangles, the sides are in proportion.

Therefore, *DC* is to *BC* as *EC* is to *AC*, or $DC/BC = EC/AC$:

$$10/15 = 12/AC$$
$$AC(10) = (15)(12) = 180$$
$$AC = 180/10 = 18$$

Therefore, side *AC* has a length of 18.

• **Example:** If you are given that *AD* is the angle bisector of vertex A in $\triangle ABC$ and *AD* intersects *BC* at E such that $(AB)(AC) = (AE)(AD)$, then prove that $\angle B \cong \angle D$.

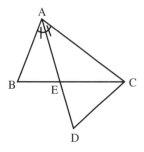

Given: $\triangle ABC$; *AD* bisects $\angle A$; $(AB)(AC) = (AE)(AD)$.

Prove: $\angle B \cong \angle D$.

Plan: Because $\angle BAE \cong \angle EAC$ and $AB/AD = AE/AC$ (by rearranging given), which reflects that two pairs of sides are in proportion, then use SAS (side-angle-side) similarity theorem to show that $\triangle BAE \sim \triangle DAC$. Because corresponding angles of similar triangles are congruent, $\angle B \cong \angle D$.

Proof:

Statements	Reasons
1. *AD* bisects *L*A.	1. Given.
2. *L*BAE ≅ *L*EAC.	2. Definition angle bisector.
3. *(AB)(AC)* = *(AE)(AD)*.	3. Given.
4. *AB/AD = AE/AC*.	4. Algebra: Divide both sides by *AC* and *AD*.
5. ΔBAE ~ ΔDAC.	5. SAS similarity theorem: If one angle of a Δ is ≅ to an angle of another Δ and corresponding sides including the *L*s are in proportion, the Δs are similar. (#2 & #4)
6. *L*B ≅ *L*D.	6. Corresponding *L*s of similar Δs are ≅.

- Two *similar triangles* can be created by drawing a line that is parallel to one of the sides of a triangle. In the figure below, ΔADC is similar to ΔAEB because the three corresponding angles are congruent. In the figure: *L*A ≅ *L*A, *L*ABE ≅ *L*ACD, and *L*AEB ≅ *L*ADC.

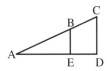

ΔADC ~ ΔAEB

The corresponding sides of similar triangles have the same *proportions:* *AC/AB = AD/AE* and *BC/AB = ED/AE*

- ***Triangle proportionality theorem*** also called the ***side-splitter theorem***: **If a line is parallel to one side of a triangle and intersects the other two sides, it divides the sides it intersects proportionally.**

In the figure below, if *DE* ‖ *BC*, then a/b = c/d. (See proof in Chapter 3, Section 3.2. *Proportional Segments.*)

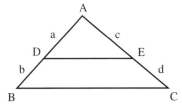

If *DE* ‖ *BC*, then a/b = c/d.
If a/b = c/d, then *DE* ‖ *BC*.
If *DE* ‖ *BC*, then a/*AB* = c/*AC* and b/*AB* = d/*AC*.

Converse to triangle proportionality theorem: If a line divides two sides of a triangle proportionally, then it is parallel to the third side.

In the figure above, if a/b = c/d, then *DE* ‖ *BC*.

Corollary to triangle proportionality theorem: If a line is parallel to one side of a triangle and intersects the other two sides, then it produces segments proportional to the side lengths.

In the figure above, if $DE \parallel BC$, then a/AB = c/AC and b/AB = d/AC.

• *Midpoints parallel length theorem*: **In a triangle, if a line connects the midpoints of two sides, then it will be parallel to the third side, and equal to half the length of the third side.**

If X is the midpoint of AB
and Y is the midpoint of AC,
then $XY \parallel BC$ and $XY = (1/2)BC$.

Prove the *midpoints parallel length theorem*:

Given: X is the midpoint of AB and Y is the midpoint of AC.

Prove: $XY \parallel BC$ and $XY = (1/2)BC$.

Plan: First show that $\triangle AXY \sim \triangle ABC$ using SAS similarity theorem, which will lead to $\angle AXY \cong \angle ABC$ and then $XY \parallel BC$ and $XY = (1/2)BC$.

Proof:

Statements	Reasons
1. X is the midpoint of AB; Y is the midpoint of AC.	1. Given.
2. $AX = (1/2)AB$; $AY = (1/2)AC$.	2. Midpoint theorem: The midpoint of a segment divides it into 2 segments each half the length of the original segment.
3. $AX/AB = 1/2$; $AY/AC = 1/2$.	3. Rearrange #2: Division.
4. $AX/AB = AY/AC$.	4. Transitive. (Shows sides have same ratio.)
5. $\angle A \cong \angle A$.	5. Identity or reflexive axiom: Self = self.
6. $\triangle AXY \sim \triangle ABC$.	6. SAS similarity theorem: If one $\angle$ of a $\triangle$ is $\cong$ to an $\angle$ of another $\triangle$ and corresponding sides that include the $\angle$s are in proportion, then the $\triangle$s are similar. (#4 & #5)
7. $\angle AXY \cong \angle ABC$.	7. Corresponding $\angle$s of similar $\triangle$s are $\cong$.
8. $XY \parallel BC$.	8. If 2 lines XY&BC are cut by transversal AB and corresponding $\angle$s are $\cong$, the lines are $\parallel$.
9. $AX/AB = XY/BC$.	9. Corresponding sides of $\sim$ $\triangle$s are in proportion.
10. $XY/BC = 1/2$.	10. Substitute #3 and #9.
11. $XY = (1/2)BC$.	11. Multiply both sides of #10 by BC.

(Also, see pages 109, 378.)

Ratios of lengths and areas

• Because *corresponding sides* of *similar triangles* have the same *proportion*, if one side is 2 times the length of its corresponding side in a similar triangle, then the other two sides of the similar triangle will be 2 times the lengths of their corresponding sides.

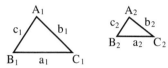

In the similar triangles, corresponding angles are congruent: $LA_1 \cong LA_2$, $LB_1 \cong LB_2$, and $LC_1 \cong LC_2$, and corresponding sides are in proportion:

$$\frac{a_2}{a_1} = \frac{b_2}{b_1} = \frac{c_2}{c_1}, \text{ or } \frac{a_1}{a_2} = \frac{b_1}{b_2} = \frac{c_1}{c_2}$$

• The *ratio of corresponding sides lengths* of similar triangles (or polygons) is called a *scale factor*. This ratio is of the measures of any two corresponding sides of two similar triangles (or polygons) and is usually given in its reduced form. If the above triangles have a scale factor of 2:1, then side a_1 is two-times larger than side a_2. The figure below depicts two similar triangles with a scale factor of 2:1.

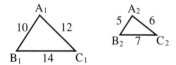

Each side of the larger triangle is two times the value of its corresponding side in the smaller triangle. The *ratios* of the corresponding sides are: $10/5 = 2/1$, $12/6 = 2/1$, and $14/7 = 2/1$

Therefore, the ratio of the corresponding sides is 2:1.

Remember, in similar triangles, the corresponding angles are congruent, $LA_1 \cong LA_2$, $LB_1 \cong LB_2$, and $LC_1 \cong LC_2$.

• The *perimeter of a triangle* is the sum of the lengths of its sides:

Perimeter of a triangle = side 1 + side 2 + side 3. It turns out that the *ratio of the perimeters of two similar triangles is the same as the ratio of corresponding sides of the two similar triangles.*

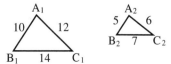

If we calculate the *perimeter* of each triangle in the figure above, by adding up their sides, we will find that the perimeters have the same ratio of 2:1 as each side.

The perimeter of $\Delta A_1B_1C_1$ is: $10 + 12 + 14 = 36$
The perimeter of $\Delta A_2B_2C_2$ is: $5 + 6 + 7 = 18$
The ratio of perimeters is: $36/18 = 2/1$

Therefore, the *ratio of the perimeters* is 2:1, which is the same as the ratio of the sides ($10/5 = 12/6 = 14/7 = 2/1$).

• *Perimeter ratio theorem*: If two similar triangles have the ratio of corresponding sides as x:y, then the *ratio of their perimeters* is x:y.

In the following similar triangles:

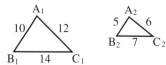

$$\frac{Perimeter\ of\ \Delta A_1B_1C_1}{Perimeter\ of\ \Delta A_2B_2C_2} = \frac{A_1B_1}{A_2B_2} = \frac{B_1C_1}{B_2C_2} = \frac{A_1C_1}{A_2C_2}$$

$$= 36/18 \qquad = 10/5 = 14/7 = 12/6 = 2/1$$

• *Segments ratio theorem*: **If two triangles are similar, then the ratio of any two corresponding segments (altitudes, medians, angle bisectors, etc.) equals the ratio of any two corresponding sides.**

In other words, the *ratio of corresponding segments of two similar triangles is the same as the ratio of corresponding sides of the two similar triangles.*

• In the following *similar triangles*, the ratio of each pair of corresponding sides is: $^{AB}/_{DE} = {}^{AC}/_{DF} = {}^{BC}/_{EF}$

Therefore:

$$\frac{Altitude\ AX}{Altitude\ DY} = \frac{AB}{DE} \qquad \frac{Angle\ Bisector\ CP}{Angle\ Bisector\ FQ} = \frac{AB}{DE} \qquad \frac{Median\ BR}{Median\ ES} = \frac{AB}{DE}$$

Altitude ⊥ base   Angle bisector   Median bisects base

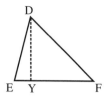 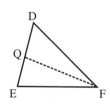 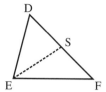

If the ratio of the corresponding *sides* in a triangle is 2:1, then the ratio of the lengths of the corresponding *altitudes* is 2:1.

If the ratio of the corresponding *sides* in a triangle is 2:1, then the ratio of the lengths of the corresponding *angle bisectors* is 2:1.

If the ratio of the corresponding *sides* in a triangle is 2:1, then the ratio of the lengths of the corresponding *medians* is 2:1.

• **Altitude/side ratio theorem**: If two triangles are similar, then the *ratio of corresponding altitudes* equals the ratio of any two *corresponding sides.*

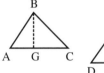

Prove this theorem:

Given: $\triangle ABC \sim \triangle DEF$; *BG* altitude of $\triangle ABC$; *EH* altitude of $\triangle DEF$.

Prove: $BG/EH = AB/DE = BC/EF = AC/DF$.

Plan: Show $\triangle ABG \sim \triangle DEH$; therefore, side ratios are in proportion so that $BG/EH = AB/DE$. Because $\triangle ABC \sim \triangle DEF$ and their side ratios are in proportion then $AB/DE = BC/EF = AC/DF$. Therefore, by transitive axiom $BG/EH = AB/DE = BC/EF = AC/DF$.

Proof:

Statements	Reasons
1. BG is altitude of $\triangle ABC$; EH is altitude of $\triangle DEF$.	1. Given.
2. $BG \perp AC$; $EH \perp DF$.	2. Definition: Altitudes $\perp$ to base.
3. $\angle BGA$ and $\angle EHD$ are right $\angle$s.	3. Perpendicular lines form right angles.
4. $\angle BGA \cong \angle EHD$.	4. Right angles are congruent.
5. $\triangle ABC \sim \triangle DEF$.	5. Given.
6. $\angle A \cong \angle D$.	6. Corresponding $\angle$s of similar $\triangle$s are $\cong$.
7. $\triangle ABG \sim \triangle DEH$.	7. AA similarity post: If two $\angle$s of one $\triangle$ are $\cong$ to $\angle$s of other $\triangle$, $\triangle$s are similar. (#6, #4)
8. $BG/EH = AB/DE$.	8. Ratio of sides of $\sim \triangle$s in proportion. (#7)
9. $AB/DE = BC/EF = AC/DF$.	9. Ratio of sides of $\sim \triangle$s in proportion. (#5)
10. $BG/EH = AB/DE$ $= BC/EF = AC/DF$.	10. Transitive axiom: Things equal to the same are $=$ to each other. (#8 & #9)

- The **area of a triangle** is: (1/2)(base)(altitude).

Area has units of length^2, such as feet^2, inches^2, meters^2, centimeters^2, or miles^2. A *side, altitude,* or *perimeter* of a triangle has units of length, such as feet, inches, meters, centimeters, or miles. Because the units of area are the square of the sides, it would follow that the **ratio of corresponding areas** *of two similar triangles is the square of the ratio of the sides of the two similar triangles.*

- **Example:** In the two triangles, each *side* of the larger triangle is two times the value of its corresponding side in the smaller triangle. Find the *ratio of the areas.*

The ratios of the corresponding sides are: $10/5 = 2/1$, $12/6 = 2/1$, and $14/7 = 2/1$. Therefore, the ratio of the corresponding sides is 2:1.

If we calculate the **area** of each triangle using (1/2)(base)(altitude):

The area of $\triangle A_1B_1C_1$ is: $(1/2)(14)(8) = 56$
The area of $\triangle A_2B_2C_2$ is: $(1/2)(7)(4) = 14$
The ratio of areas is: $56/14 = 4/1$ and $4/1 = 2^2/1^2$

Therefore, the ratio of the areas is 4:1, which is the *square of* the ratio of the sides, which is 2:1.

• *Area ratio theorem*: **If two similar triangles have the ratio of corresponding sides as x:y, then the ratio of their areas is x^2:y^2.**

• **Example:** You have been contracted by your town to build two triangle-shaped parks. The first should have equal side lengths of 600 feet each, and the second should be 2/3 the size of the first. (a) How much area will you need to cover with grass and trees for each park? (b) What is the ratio of the areas? (c) What is the perimeter of each park? (d) What is the ratio of the perimeters?

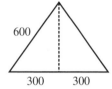 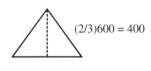

(a) To find the area of the larger triangle, we know the formula for area is (1/2)(altitude)(base). The base is 600 feet. We can find the length of the altitude because it splits the equilateral triangle into two 30:60:90 triangles. In this case, the larger equilateral triangle has a hypotenuse of 600 and a short leg of 300. The long leg of each 30:60:90 triangle is the altitude of the equilateral triangle.

We can find the long leg (or altitude) using the Pythagorean Theorem, $x^2 + y^2 = z^2$, or $300^2 + (\text{altitude})^2 = 600^2$

Using Pythagorean Theorem:

(altitude)^2 = $600^2 - 300^2$ = 270,000, or (altitude)^2 = 270,000

Take the square root of both sides:

altitude = $\sqrt{(3)(300)(300)}$ = 300$\sqrt{3}$ ≈ 519.615...

Therefore, the area of the large triangle = (1/2)(altitude)(base)

= (1/2)(600)(300$\sqrt{3}$) = 90,000$\sqrt{3}$ ≈ 155,884.57 square feet.

To find the area for the smaller triangle, we can use the *ratio of areas*.

The ratio of the areas is the square of the ratio of the sides:

400^2:600^2 = 160,000:360,000 = 160,000/360,000 = 16/36 = 4/9

Or we could have reduced the sides first: 400/600 = 4/6 = 2/3.

Then square: 2^2:3^2 = 4/9

Therefore, the ratio of the areas is 4:9 or 4/9.

We can use the ratio of areas to find the area of the smaller triangle.

   Area small triangle / Area big triangle =
   Area small triangle / $90,000\sqrt{3}$ = 4/9

Solve this proportion for the area of the small triangle by multiplying both sides by $90,000\sqrt{3}$:

   Area small triangle = $(90,000\sqrt{3})4/9 = 40,000\sqrt{3} \approx 69,282.03$

Therefore, the area of the large triangle is $90,000\sqrt{3} \approx 155,884.57$ square feet, and the area of the small triangle is $40,000\sqrt{3} \approx 69,282.03$ square feet.

(b) We determined the ratio of the areas in part (a) as the square of the ratio of the sides, $400^2:600^2 = 160,000/360,000 = 16/36 = 4/9$

Therefore, the ratio of the areas is 4:9 or 4/9.

Remember the *area ratio theorem*: If two similar triangles have a ratio of corresponding sides of x:y, then the ratio of their areas is $x^2:y^2$.

(c) The perimeter of each park is the sum of its sides.

   For the large triangle $600 + 600 + 600 = 3(600) = 1,800$ feet.
   For the small triangle $400 + 400 + 400 = 3(400) = 1,200$ feet.

(d) The ratio of the perimeters of the small triangle to the large triangle is: $1,200/1,800 = 12/18 = 2/3$, which is the same as the ratio of the sides. Remember the *perimeter ratio theorem*, which states that if two similar triangles have a corresponding sides ratio of x:y, then the *ratio of their perimeters* is x:y.

4.7 Similar Right Triangles

• When working with *similar right triangles*, the definitions and principles of *similar triangles* hold true. Especially important definitions are:

Corresponding angles of similar triangles are congruent, and corresponding sides of similar triangles are in proportion (side lengths have the same ratio).

• An extension of the *AA similarity triangle postulate* that can be used to determine if two **right triangles** are similar is the following theorem.

Acute angle similarity theorem: **Two right triangles are similar if an acute angle of one triangle is congruent to an acute angle of the other.**

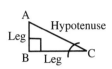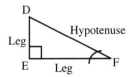

We know that right angles B and E are congruent. Therefore if one other pair of corresponding angles are congruent (either $\angle A$ and $\angle D$ or $\angle C$ and $\angle F$), then the triangles are similar.

The *acute angle similarity theorem* is the same as having two congruent angles, which is what is required by the *AA similarity triangle postulate*: If *two* angles of one triangle are congruent to *two* angles of another triangle, then the triangles are *similar*.

Review of Geometric Mean

• The geometric mean is of use in the proportions of similar right triangles that can be created by drawing an altitude to the hypotenuse of a right triangle. In Section 3.1 *Ratios and Proportions*, we learned that in a proportion, such as a:b::b:c, if the two means terms (b and b) are the same, then b is the **geometric mean**, also called **mean proportional**, between the first and fourth terms, a and c.

In the proportion a:b::b:c or equivalently a/b = b/c, term b is the *geometric mean* between a and c. Cross-multiplying gives:

$b^2 = (a)(c)$, or $b = \sqrt{ac}$

The geometric mean of two numbers is the square root of their product.

Altitude From the Right Angle to the Hypotenuse of a Right Triangle

• The *altitude from the right angle to the hypotenuse* of a right triangle splits the right triangle into two *similar triangles*, each of which is also similar to the original triangle. This can be written as a theorem:

Altitude to hypotenuse theorem: **The altitude from the right angle to the hypotenuse of a right triangle divides it into two triangles that are similar to the original triangle and similar to each other**. (See informal proof in example below.)

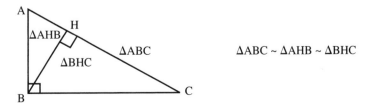

Because each of these triangles is similar to each other, the *ratios of each pair of corresponding sides is equal.*

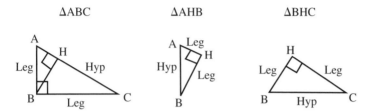

We can write the ratios of corresponding sides for each triangle pair:

Hyp/SLeg = Hyp/SLeg; Hyp/LLeg = Hyp/LLeg; SLeg/LLeg = SLeg/LLeg

$\triangle ABC \sim \triangle AHB$: <u>$AC/AB = AB/AH$</u>; $AC/BC = AB/HB$; $AB/BC = AH/HB$

$\triangle AHB \sim \triangle BHC$: $AB/AH = BC/BH$; $AB/BH = BC/CH$; <u>$AH/BH = BH/CH$</u>

$\triangle ABC \sim \triangle BHC$: $AC/AB = BC/BH$; <u>$AC/BC = BC/HC$</u>; $AB/BC = BH/CH$

From the side ratios, we can see that one of the proportions for each pair of similar triangles contains a **geometric mean,** also called *mean proportional,* (underlined).

• The three similar triangles and their *geometric mean*-containing proportions are listed below. There are two corollaries of the *altitude to hypotenuse theorem* that relate the parts of these similar right triangles to each other. The relations between the parts of these similar right triangles, (which are constructed by drawing the altitude to the hypotenuse of any right triangle), can by useful if you need to determine particular parts of the primary triangle.

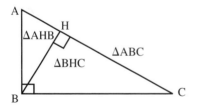

$\triangle ABC \sim \triangle BHC$ contains a geometric mean: $AC/BC = BC/HC$

$\triangle ABC \sim \triangle AHB$ contains a geometric mean: $AC/AB = AB/AH$

$\triangle AHB \sim \triangle BHC$ contains a geometric mean: $AH/BH = BH/CH$

• *Corollary to altitude to hypotenuse theorem-leg as geometric mean*:
In a right triangle (ΔABC), the length of either leg (*AB* or *BC*) is the
geometric mean (*mean proportional*) between the length of the hypotenuse
(*AC*) and the length of the projection of that leg onto the hypotenuse
(between the *altitude* line and each vertex) (*AH* or *HC*).

In other words, **each *leg* is the *geometric mean* between the hypotenuse
and the projection of that leg onto the hypotenuse.**

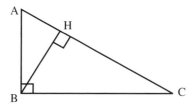

$AC/AB = AB/AH$,   *AB* is the geometric mean between *AC* and *AH*.

$$AB = \sqrt{AC \times AH}$$

$AC/BC = BC/HC$,   *BC* is the geometric mean between *AC* and *HC*.

$$BC = \sqrt{AC \times HC}$$

This corollary is true because corresponding sides of similar triangles are
in proportion.

• *Corollary to altitude to hypotenuse theorem-altitude as geometric
mean*: In a right triangle (ΔABC), the length of the *altitude* (*BH*) drawn
from the right angle to the hypotenuse (*AC*), is the *geometric mean*
(*mean proportional*) between the resulting segments on the hypotenuse
(*AH* and *CH*).

In other words, **the altitude to the hypotenuse is the *geometric mean*
between the resulting segments on the hypotenuse.**

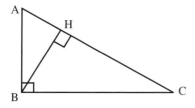

$AH/BH = BH/CH$,   *BH* is the geometric mean between *AH* and *CH*.

$$BH = \sqrt{AH \times CH}$$

• **Example:** Find c, h, and a.

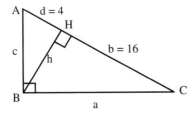

To find h, use the *corollary to altitude to hypotenuse theorem-altitude as geometric mean*, which states that *in a right triangle (ΔABC), the length of the altitude (BH) is the geometric mean between the resulting segments on the hypotenuse (AH and CH).*

AH and *CH* are 4 and 16, respectively. The proportion is:

$AH/BH = BH/CH = 4/h = h/16$

Cross-multiply:

$h^2 = 64$

$h = \sqrt{64} = 8$

To find c, use the *corollary to altitude to hypotenuse theorem-leg as geometric mean*, which states that *in a right triangle (ΔABC), the length of either leg (AB or BC) is the geometric mean between the length of the hypotenuse (AC) and the length of the projection of that leg onto the hypotenuse (AH or HC).*

$AC = AH + CH$; where $AH = 4$ and $CH = 16$. The proportion is:

$AC/AB = AB/AH = (4+16)/c = c/4$

$c^2 = (20)(4) = 80$

$c = \sqrt{80} = \sqrt{4 \times 4 \times 5} = 4\sqrt{5}$

To find a, use the *corollary to altitude to hypotenuse theorem-leg as geometric mean*, which states that *in a right triangle (ΔABC), the length of either leg (AB or BC) is the geometric mean between the length of the hypotenuse (AC) and the length of the projection of that leg onto the hypotenuse (AH or HC).*

$AC = AH + CH$; where $AH = 4$ and $CH = 16$. The proportion is:

$AC/BC = BC/HC = (4+16)/a = a/16$

$a^2 = (20)(16) = 320$

$a = \sqrt{320} = \sqrt{8 \times 8 \times 5} = 8\sqrt{5}$

Therefore, $c = 4\sqrt{5}$, $h = 8$, and $a = 8\sqrt{5}$.

• **Example:** Write an informal proof stating reasons why the ***altitude to hypotenuse theorem*** is true. This theorem states that the altitude from the right angle to the hypotenuse of a *right triangle* divides it into two triangles that are similar to the original triangle and similar to each other.

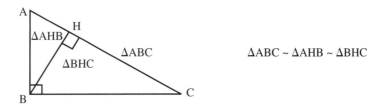

Given: △ABC is a right triangle; *BH* is the altitude.

Prove: △ABC ~ △AHB ~ △BHC.

Informal proof:

Because △AHB and △ABC have a common (congruent) acute angle *L*A, and they are both right triangles, then using the acute angle similarity theorem (or the AA similarity triangle postulate): △ABC ~ △AHB.

Because △BHC and △ABC have a common (congruent) acute angle *L*C and they are both right triangles, then using the acute angle similarity theorem (or the AA similarity triangle postulate): △ABC ~ △BHC.

Because △ABC ~ △AHB and △ABC ~ △BHC, and because two triangles similar to a third triangle (△ABC) are similar to each other, then △AHB ~ △BHC.

Therefore, △ABC ~ △AHB ~ △BHC.

4.8 Right Triangles: Pythagorean Theorem and 30°:60°:90° and 45°:45°:90° Triangles

• A ***right triangle***, is defined as a planar triangle having one of its interior *angles* as a *right angle* measuring 90°. ***Right triangles*** consist of one right (90°) angle and two acute (<90°) angles that sum to 90°, so that the total sum of the angles is 180°.

• In a ***right triangle***, the side opposite to the right angle is called the ***hypotenuse***, and the two sides that meet to form the right angle are called ***legs***. The hypotenuse is always the longest side.

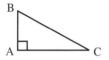

$m\angle a < 90°$, $m\angle b = 90°$, $m\angle c < 90°$
□ denotes the 90° angle.

• The *acute angles of a right triangle are complementary* and therefore sum to 90°. Remember, the sum of the three angles in any planar triangle is 180°, so if one angle is 90°, then the other two must sum to 90°.

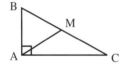

Angle B + Angle C = 90° because Angle A = 90°, and Angle A + Angle B + Angle C must = 180°.

• A triangle can have no more than one right angle (because the total sum of the angles is 180°).

• If a *median* is drawn from the right angle to the hypotenuse, it (the median) will be half the length of the hypotenuse.

Remember, a *median* of a triangle is a line segment extended from any of the three angle vertexes to the *center or midpoint* of the side opposite the angle. A median *bisects* the side to which it is drawn.

 Median bisects side.

Right triangle median theorem: In a *right triangle*, the *median from the right angle* to the hypotenuse is *one-half the length of the hypotenuse*.

Median $AM = (1/2)BC$

• **Example:** You are buying a house with a triangular yard, ΔABC, that backs up to a perfect right corner of the L-shaped house. Before moving in, you want to order fencing across *BC* (in the drawing) and there is a pole M that is exactly between B and C. The builder told you that the distance from the inside corner to the pole, *AM*, is 20 feet, but he doesn't know the length of *BC* across the back of the yard. How can you figure out length *BC* before moving in and measuring it?

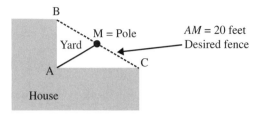

You remember from your geometry class that *AM* forms the median of $\triangle$ABC, and in a right triangle, the median (*AM*) to the hypotenuse is one-half the hypotenuse (*BC*) length.

If the median (*AM*) from the right angle to the hypotenuse is 20 feet, then to find the length of the hypotenuse (*BC*) we can use:

$AM = (1/2)BC$

$BC = (2)AM = (2)(20 \text{ feet}) = 40 \text{ feet}$

Therefore, the length of the fencing you need to order is 40 feet.

• ***Hypotenuse midpoint theorem***: In a right triangle, the midpoint of the hypotenuse is equidistant from the three vertices.

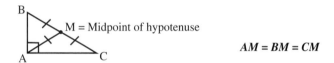

• **Example:** In the drawing of right triangle $\triangle$ABC, *CD* is the *altitude* to hypotenuse *AB*, *CE* is the *angle bisector* of $\angle$ACB, and *CF* is the *median* to hypotenuse *AB*. Prove that $\angle$DCE $\cong$ $\angle$ECF.

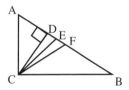

Given: Right triangle $\triangle$ABC; *CD* is *altitude* to *AB*; *CE* is *angle bisector* of $\angle$ACB; and *CF* is *median* to *AB*.

Prove: $\angle$DCE $\cong$ $\angle$ECF.

Proof:

Statements	Reasons
1. Rt. ΔABC; *CD* alt to *AB*; *CE* angle bisector of ∠ACB; *CF* med to *AB*.	1. Given.
2. *CF* ≅ (1/2)*AB*.	2. Median to hypotenuse of Rt. Δ is (1/2)hypotenuse.
3. *CF* ≅ *FB*.	3. Definition of median.
4. ΔCFB is isosceles triangle.	4. Definition of isosceles Δ: 2 ≅ sides.
5. ∠BCF ≅ ∠B.	5. Δ with 2 ≅ sides has ≅ ∠s opposite sides.
6. ∠A is complementary to ∠ACD.	6. ΔACD is a right Δ.
7. ∠A is complementary to ∠B.	7. ΔABC is a right Δ.
8. ∠ACD ≅ ∠B.	8. Complements to same ∠ are ≅. (#6 and #7)
9. ∠ACD ≅ ∠BCF.	9. Transitive/Substitution of #5 & #8.
10. ∠ACE ≅ ∠BCE.	10. Given *CE* is bisector of ∠ACB.
11. ∠DCE ≅ ∠FCE	11. Subtract angles in #9 from #10.

Pythagorean Theorem

• *Pythagorean Theorem*: **In a right triangle, the square of the length of the hypotenuse is equal to the sum of the squares of the lengths of the legs.**

$$(\text{Leg})^2 + (\text{Leg})^2 = (\text{Hypotenuse})^2$$

This is called the *Pythagorean Theorem and it only applies to right triangles*. If the lengths of the legs are x and y, and the length of the hypotenuse is z, the Pythagorean Theorem can be written: $x^2 + y^2 = z^2$.

• There are many *proofs of the Pythagorean Theorem*. Following is a proof based on the ratios of the sides of a right triangle with an altitude line drawn from the right angle to the hypotenuse (discussed in the preceding Section 4.7. *Similar Right Triangles*).

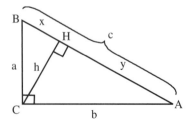

Given: $\triangle ABC$ with right angle $\angle BCA$, leg lengths a and b, and hypotenuse length c, where c = x + y.

Prove: $a^2 + b^2 = c^2$.

Plan: Draw altitude *CH* and use corollary to altitude to hypotenuse theorem, *in a right triangle, each leg is the geometric mean between the hypotenuse and the projection of that leg onto the hypotenuse*, to show that c/a = a/x and c/b = b/y. Then add these proportions and use the segment addition postulate for x + y = c, to obtain $a^2 + b^2 = c^2$.

Proof:

Statements	Reasons
1. $\triangle ABC$: Rt. $\angle C$, legs a & b, hyp. c.	1. Given.
2. Draw altitude *CH*, which is a perpendicular from $\angle C$ to *AB*.	2. Perpendicular point to line postulate: Only one $\perp$ line can be drawn from any point *not* on a line to that line.
3. c/a = a/x and c/b = b/y.	3. Each leg of a Rt. $\triangle$ is the geometric mean between the hypotenuse and the leg's projection on the hypotenuse.
4. $a^2 = cx$ and $b^2 = cy$.	4. Cross-multiply each proportion in #3.
5. $a^2 + b^2 = cx + cy$.	5. Addition of equations in #4.
6. $a^2 + b^2 = c(x + y)$.	6. Factor equation in #5.
7. c = x + y.	7. Given.
8. $a^2 + b^2 = c^2$.	8. Substitution of #6 and #7.

• **Converse of Pythagorean Theorem**: If a triangle has side lengths x, y, and z, where z is the longest side and x, y, and z satisfy the equation $x^2 + y^2 = z^2$, then the triangle is a right triangle with z as its hypotenuse. In other words, if the sum of the squares of two sides of a triangle is equal to the square of the third side, then the triangle is a right triangle.

If $x^2 + y^2 = z^2$, then the triangle is a right triangle.

• **Areas of squares surrounding right triangle**: The Pythagorean Theorem tells us that if squares are drawn onto each of the three sides of a right triangle, *the sum of the areas of the squares on the two legs is equal to the area of the square on the hypotenuse.*

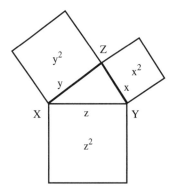

If x and y are the leg lengths and z is the hypotenuse length then: $x^2 + y^2 = z^2$.

- **Example:** In right $\triangle XYZ$ (above) with z as the hypotenuse and x and y as legs, if x = 3 and y = 4, what is the area of each of the three squares along the sides x, y, and z?

Area $x^2 = 3^2 = 9$
Area $y^2 = 4^2 = 16$

Using the Pythagorean Theorem, $x^2 + y^2 = z^2$:

$z^2 = 9 + 16 = 25$, or $z = \sqrt{25} = 5$

Therefore, the sides x, y, and z of this right triangle are 3, 4, and 5, respectively. The 3:4:5 right triangle is a right triangle Pythagorean *triplet* and you should become familiar with it.

- **Example:** In right $\triangle XYZ$ (above) with z as the hypotenuse and x and y as legs, if the area $x^2 = 25$ and the area $y^2 = 144$, what is x, y, and z?

First, x and y are the square roots of 25 and 144, respectively:

$x = \sqrt{25} = 5$

$y = \sqrt{144} = 12$

We can find z using the Pythagorean Theorem, $x^2 + y^2 = z^2$, to obtain z^2 and then z.

$x^2 + y^2 = z^2$, or, $25 + 144 = z^2$

$z = \sqrt{169} = 13$

Therefore, the sides x, y, and z of this right triangle are 5, 12, and 13, respectively. The 5:12:13 right triangle is called a right triangle Pythagorean *triplet*, and you should become familiar with it.

- **Example:** Find the shorter leg, x, of the right triangle.

17
x
15

Using the Pythagorean Theorem: $(\text{Leg})^2 + (\text{Leg})^2 = (\text{Hypotenuse})^2$

$x^2 + (15)^2 = (17)^2$

$x^2 + 225 = 289$

$x^2 = 289 - 225 = 64$

$x = \sqrt{64} = 8$, which is the shorter leg.

This triangle with side lengths 8:15:17 is a right triangle Pythagorean triplet and its sides satisfy the equation $x^2 + y^2 = z^2$, or $8^2 + 15^2 = 17^2$.

- **Example:** Find the hypotenuse, x, of the right triangle.

Using the Pythagorean Theorem: $(\text{Leg})^2 + (\text{Leg})^2 = (\text{Hypotenuse})^2$

$(5)^2 + (12)^2 = (\text{Hypotenuse})^2 = 25 + 144 = (\text{Hypotenuse})^2$

$169 = (\text{Hypotenuse})^2$

Hypotenuse $= \sqrt{169} = 13$

Therefore, the hypotenuse x is 13. This triangle having side lengths 5:12:13 is a right triangle Pythagorean triplet because 5, 12, and 13 satisfy the Pythagorean Theorem: $x^2 + y^2 = z^2$, or $5^2 + 12^2 = 13^2$.

- **Example:** Find the altitude, y, to the base of the *isosceles triangle*.

Because the two non-base side lengths of an isosceles triangle are by definition equal, then $z = 5$.

Also, because the altitude y to the base of an isosceles triangle bisects the base, then we can determine the altitude y by splitting the isosceles triangle into two right triangles.

Use the Pythagorean Theorem to find y for one of the right triangles:

$(\text{Leg})^2 + (\text{Leg})^2 = (\text{Hypotenuse})^2$, or, $(x/2)^2 + y^2 = z^2$

$(6/2)^2 + y^2 = 5^2$

$3^2 + y^2 = 5^2$

$9 + y^2 = 25$

$y^2 = 25 - 9 = 16$

$y = \sqrt{16} = 4$

Therefore, the altitude y to the base of this isosceles triangle is 4.

Note that the two right triangles split from the isosceles triangle in this example are 3:4:5 right triangle Pythagorean triplets.

• A *modification of the Pythagorean Theorem* allows us to determine whether a triangle is an *acute triangle* or an *obtuse triangle*.

Acute Pythagorean-variation theorem: If a triangle has side lengths x, y, and z, where z is the longest side, then the triangle is *acute* if $z^2 < x^2 + y^2$.

Obtuse Pythagorean-variation theorem: If a triangle has side lengths x, y, and z, where z is the longest side, then the triangle is *obtuse* if $z^2 > x^2 + y^2$.

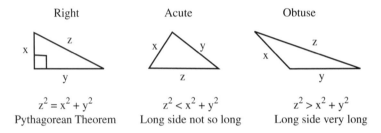

Right	Acute	Obtuse
$z^2 = x^2 + y^2$	$z^2 < x^2 + y^2$	$z^2 > x^2 + y^2$
Pythagorean Theorem	Long side not so long	Long side very long

If $z^2 \neq x^2 + y^2$, then the triangle is *not* a right triangle and must be either acute or obtuse.

• **Example:** Prove the *acute Pythagorean-variation theorem*: If a triangle has side lengths x, y, and z, where z is the longest side, then the triangle is *acute* if $z^2 < x^2 + y^2$.

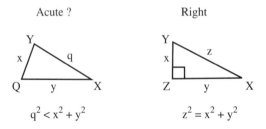

Acute ?	Right
$q^2 < x^2 + y^2$	$z^2 = x^2 + y^2$

Given: ΔXYQ having q as the longest side satisfies $q^2 < x^2 + y^2$; ΔXYZ is a right triangle with ∠Z as right angle.

Prove: ΔXYQ is an acute triangle.

Plan: Show that q < z and therefore $m∠Q < m∠Z = 90°$ using SSS *inequality* theorem. It follows that ΔXYQ is an acute triangle.

Proof:

Statements	Reasons
1. $\triangle XYQ$ with q as longest side.	1. Given.
2. $q^2 < x^2 + y^2$.	2. Given.
3. $\triangle XYZ$ is Rt. $\triangle$ with $\angle Z$ Rt $\angle$.	3. Given.
4. $z^2 = x^2 + y^2$.	4. Pythagorean Theorem.
5. $q^2 < z^2$; therefore, $q < z$.	5. Substitution of #2 & #4; $q^2 < x^2 + y^2 = z^2$.
6. $m\angle Q < m\angle Z$.	6. SSS *inequality* theorem: If two corresponding sides of 2 $\triangle$s are $\cong$, $\triangle$ with greater included $\angle$ has longer third side opposite the $\angle$.
7. $m\angle Z = 90°$.	7. Definition of right angle.
8. $m\angle Q < 90°$.	8. Substitute #6 and #7; $m\angle Q < m\angle Z = 90°$.
9. $\triangle XYQ$ is an acute triangle.	9. Definition of acute $\triangle$: $\angle$s $< 90°$.

• **Example:** Prove the *obtuse Pythagorean-variation theorem*: If a triangle has side lengths x, y, and z, where z is the longest side, then the triangle is **obtuse** if $z^2 > x^2 + y^2$.

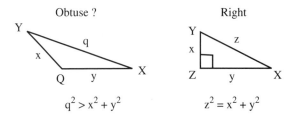

Given: $\triangle XYQ$, having q as the longest side, satisfies $q^2 > x^2 + y^2$; $\triangle XYZ$ is a right triangle with $\angle Z$ as right angle.

Prove: $\triangle XYQ$ is an obtuse triangle.

Plan: Show that $q > z$ and therefore $m\angle Q > m\angle Z = 90°$ using SSS *inequality* theorem. It follows that $\triangle XYQ$ is an obtuse triangle.

Proof:

Statements	Reasons
1. $\triangle XYQ$ with q as longest side.	1. Given.
2. $q^2 > x^2 + y^2$.	2. Given.
3. $\triangle XYZ$ is Rt. $\triangle$ with $\angle Z$ Rt $\angle$.	3. Given.
4. $z^2 = x^2 + y^2$.	4. Pythagorean Theorem.
5. $q^2 > z^2$; therefore, $q > z$.	5. Substitution of #2 and #4; $q^2 > x^2 + y^2 = z^2$.
6. $m\angle Q > m\angle Z$.	6. SSS *inequality* theorem: If two corresponding sides of 2 $\triangle$s are $\cong$, $\triangle$ with greater included $\angle$ has longer third side opposite the $\angle$.
7. $m\angle Z = 90°$.	7. Definition of right angle.
8. $m\angle Q > 90°$.	8. Substitute #6 and #7; $m\angle Q > m\angle Z = 90°$.
9. $\triangle XYQ$ is an obtuse triangle.	9. Definition of obtuse $\triangle$: one $\angle > 90°$.

- **Example:** Are the following right, acute, or obtuse triangles?
(a) If x = 3, y = 5, and z = 9. (b) If x = 10, y = 24, and z = 26.
(c) If x = 20, y = 50, and z = 100.

(a) If x = 3, y = 5, and z = 9, calculate $z^2 = x^2 + y^2$:
Does $9^2 = 3^2 + 5^2$? $\longrightarrow$ $81 = 9 + 25 = 34$ $\longrightarrow$ $81 \neq 34$
$81 > 34$; therefore, $z^2 > x^2 + y^2$ and this is an *obtuse* triangle.

(b) If x = 10, y = 24, and z = 26, calculate $z^2 = x^2 + y^2$:
Does $26^2 = 10^2 + 24^2$? $\longrightarrow$ $676 = 100 + 576$ $\longrightarrow$ $676 = 676$
$676 = 676$; therefore, $z^2 = x^2 + y^2$ and this is an *right* triangle.

(c) If x = 40, y = 50, and z = 60, calculate $z^2 = x^2 + y^2$:
Does $60^2 = 40^2 + 50^2$? $\longrightarrow$ $3,600 = 1,600 + 2,500$ $\longrightarrow$ $3,600 \neq 4,100$
$3,600 < 4,100$; therefore, $z^2 < x^2 + y^2$ and this is an *acute* triangle.

30°:60°:90° and 45°:45°:90° Triangles

- There are a few common *right triangles* that are important for you to recognize. These include the **30°:60°:90°** *right triangle* and the **45°:45°:90°** (*right isosceles*) *triangle*. The properties of these triangles are particularly useful in trigonometry and in the trigonometric functions, which are briefly described in this chapter and described in detail in *Master Math: Trigonometry*.

45°:45°:90° (Right Isosceles Triangle)

- The **45°:45°:90°** (*right isosceles triangle*) has two equal angles, two equal sides, and a right angle.

The ratio of the sides, leg : leg : hypotenuse, is: $x : x : x\sqrt{2}$, or equivalently $1 : 1 : \sqrt{2}$, or equivalently $1/\sqrt{2} : 1/\sqrt{2} : 1$.

Note that $\sqrt{2} = 1.41421356...$ is an irrational number and possesses endless non-repeating digits to the right of the decimal point. To assure accuracy, especially during calculations, it is generally left in its square-root form.

• The 45°:45°:90° right triangle forms half of a *square* with the hypotenuse of the triangle as the diagonal of the square. We can use this along with the Pythagorean Theorem to *derive the triangle side ratios*:

z = hypotenuse of triangle and diagonal of square.
x = leg of triangle and side of square.
y = leg of triangle and side of square, and x = y.

Because $x = y$ in a *square*, the Pythagorean Theorem, $(\text{Leg})^2 + (\text{Leg})^2 = (\text{Hypotenuse})^2$, or $x^2 + y^2 = z^2$, becomes: $2(x)^2 = z^2$

$z = \sqrt{2x^2} = x\sqrt{2}$

This results in side ratios: $x : x : x\sqrt{2}$ for x, y, and z, respectively.

Principles That Apply To a 45°:45°:90° (*Right Isosceles*) Triangle

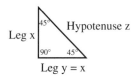

• *Isosceles right triangle theorem*: In an *isosceles right triangle*, the length of the hypotenuse z equals the length of a leg, x, multiplied by $\sqrt{2}$: $z = x\sqrt{2}$.

• *Corollary to isosceles right triangle theorem*: The length of the diagonal d in a *square* equals the length of a side multiplied by $\sqrt{2}$: diagonal $d = (\text{side})\sqrt{2}$.

Side

• *Principle*: The measures of each acute angle in an *isosceles right triangle* is 45°.

• *Principle*: The length of a leg x opposite a 45° angle equals one-half the length of the hypotenuse z multiplied by $\sqrt{2}$: $x = (1/2)z\sqrt{2}$.

• **Example:** Find all side lengths in the right triangle below.

Begin by determining the unknown acute angle α using the fact that the angles in a triangle sum to 180°: $45° + 90° + \alpha = 180°$

$\alpha = 180° - 45° - 90° = 45°$

A right triangle with two 45° angles is an isosceles right triangle, having two legs of the same length, and the ratio of its sides are: $x : x : x\sqrt{2}$.

Therefore, both legs have length 5 and the hypotenuse has length $5\sqrt{2}$ resulting in sides of $x = 5$, $y = 5$, and $z = 5\sqrt{2}$.

An alternative method of determining the hypotenuse, once we know the triangle is a right isosceles triangle with two 45° angles and its two legs each having a length of 5, is to use the Pythagorean Theorem: $(\text{Leg})^2 + (\text{Leg})^2 = (\text{Hypotenuse})^2$

$z^2 = 5^2 + 5^2 = 25 + 25$

$z^2 = 50$, or $z = \sqrt{5 \times 5 \times 2} = 5\sqrt{2}$

30°:60°:90° Triangles

• The **30°:60°:90° triangles** have a right angle, two acute angles, 30° and 60°, and three unequal sides.

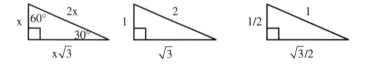

The ratio of the sides, leg : leg : hypotenuse, is: $x : x\sqrt{3} : 2x$, or equivalently $1 : \sqrt{3} : 2$, or equivalently $1/2 : \sqrt{3}/2 : 1$.

The hypotenuse is the longest side opposite the 90° angle and has a length of 2x.

The short leg is opposite the 30° angle and has a length of x.

The long leg is opposite the 60° angle and has a length of $x\sqrt{3}$.

• The 30°:60°:90° triangle forms half of an ***equilateral triangle*** with the *long leg as the altitude*. We can use this along with the Pythagorean Theorem to derive the side ratios, which is shown after the principles.

Principles That Apply to a 30°:60°:90° Triangle

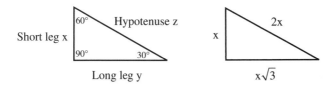

• ***30°:60°:90° triangle theorem***: In a 30°:60°:90° triangle, the hypotenuse is twice as long as the shorter leg, and the longer leg is $\sqrt{3}$ times longer than the shorter leg.

Hypotenuse $= 2 \times$ short leg

Long leg $= \sqrt{3} \times$ short leg

• ***Principle***: The length of *short leg* x opposite the 30° angle equals one-half the length of the hypotenuse z: x = z/2, (or z = 2x)

• ***Principle***: The length of *long leg* y opposite the 60° angle equals one-half the length of the hypotenuse z times $\sqrt{3}$: y = $(1/2)z\sqrt{3}$. (Substituting for z: y = $(1/2)2x\sqrt{3}$ = $x\sqrt{3}$.)

• ***Principle***: The length of *long leg* y opposite the 60° angle equals the length of short leg x opposite the 30° angle times $\sqrt{3}$: y = $x\sqrt{3}$.

• ***Corollary to 30°:60°:90° triangle theorem***: The length of the *altitude of an equilateral triangle* equals *one-half* the length of a side of the equilateral triangle times $\sqrt{3}$. For an equilateral triangle:

altitude = $(1/2)(\text{side}) \sqrt{3}$, or for 30°:60°:90° triangle y = $(1/2)z\sqrt{3}$.

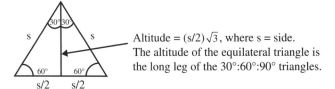

Altitude = $(s/2)\sqrt{3}$, where s = side.
The altitude of the equilateral triangle is
the long leg of the 30°:60°:90° triangles.

Note that $\sqrt{3}$ = 1.7320508... is an irrational number having endless non-repeating digits, and is left in its square-root form to assure accuracy.

The 30°:60°:90° Triangle and the Equilateral Triangle

• A unique property of the *30°:60°:90° triangle* is that it forms half of an *equilateral triangle* with the long leg as the altitude. In other words, an equilateral triangle contains two 30°:60°:90° triangles.

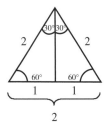

We can *derive the side ratios* of a 30°:60°:90° triangle by applying the Pythagorean Theorem to one of the right triangles in the equilateral triangle. For the derivation, let the sides of the equilateral triangle be 2.

Using the Pythagorean Theorem, $(Leg)^2 + (Leg)^2 = (Hypotenuse)^2$, and the above figure, we can determine the altitude of the equilateral triangle, which is the long leg of the 30:60:90 triangle:

$(Short\ Leg)^2 + (Long\ Leg)^2 = (Hypotenuse)^2$
$1^2 + (Long\ Leg)^2 = 2^2$
$1 + (Long\ Leg)^2 = 4$
$(Long\ Leg)^2 = 4 - 1 = 3$
$Long\ Leg = \sqrt{3}$

By choosing a hypotenuse of 2 and short leg of 1, we calculated the long leg as $\sqrt{3}$. This results in the side ratios of a 30°:60°:90° triangle of:

$1 : \sqrt{3} : 2$

• If we know the *perimeter* of an equilateral triangle, we can find its *altitude* using the side ratios of 30°:60°:90° triangles, which are:

$x : x\sqrt{3} : 2x$

• **Example:** Given a perimeter of 21, find the *altitude* of an equilateral triangle.

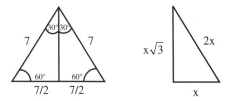

Because each side of an equilateral triangle has the same length, a perimeter of 21 divided by 3 sides equals side lengths of 7. Therefore, the hypotenuse of each of the two 30°:60°:90° triangles inside the equilateral triangle is 7. We can use the side ratio of the 30°:60°:90° triangles, $x : x\sqrt{3} : 2x$:

  Hypotenuse $= 7 = 2x$
  Rearrange: $x = 7/2$

The altitude of the equilateral triangle is the long leg opposite the 60° angle, which has a length: $x\sqrt{3} = 7/2\sqrt{3}$

Therefore, the altitude of this equilateral triangle is: $7/2\sqrt{3}$.

We can check this result using the Pythagorean Theorem:

$leg^2 + leg^2 = hypotenuse^2$. If the hypotenuse of the 30°:60°:90° triangle is 7 and the short leg is 7/2, does the long leg, or altitude of the equilateral triangle, equal $7/2\sqrt{3}$?

  $(7/2\sqrt{3})^2 + (7/2)^2 = (7)^2$, or equivalently $(3)49/4 + 49/4 = 49$, or

  $36.75 + 12.25 = 49$, or $49 = 49$

• **Example:** Find the legs if the hypotenuse of a 30:60:90 triangle is: (a) 12 or (b) 50.

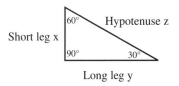

To find the legs, we can use the side ratios of a 30:60:90 triangle $(x : x\sqrt{3} : 2x)$ and the principles:

Hypotenuse $= z = 2x$.

Short leg $= x = z/2$. The length of *short leg* x opposite the 30° angle equals one-half the length of the hypotenuse z.

Long leg $= y = (1/2)z\sqrt{3} = x\sqrt{3}$. The length of *long leg* y opposite the 60° angle equals one-half the length of the hypotenuse z times $\sqrt{3}$.

(a) Hypotenuse $z = 12$:

  $x = 12/2 = 6$

  $y = (1/2)(12)\sqrt{3} = 6\sqrt{3}$

(b) Hypotenuse $z = 50$:

  $x = 50/2 = 25$

  $y = (1/2)(50)\sqrt{3} = 25\sqrt{3}$

We can check this result using the Pythagorean Theorem:

  $\text{leg}^2 + \text{leg}^2 = \text{hypotenuse}^2$

  (a) $6^2 + (6\sqrt{3})^2 = 12^2$, or $36 + 108 = 144$, or $144 = 144$

  (b) $25^2 + (25\sqrt{3})^2 = 50^2$, or $625 + 1{,}875 = 2{,}500$, or $2{,}500 = 2{,}500$

• **Example:** Given a triangle having one 60 degree angle in a circle with a diameter of 40, what is the length of chords AB and BC?

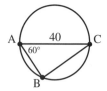 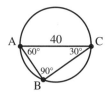

An angle inscribed in a semicircle will always have a measure of 90°. Because triangle $\triangle ABC$ is inscribed in a semi-circle with one of its sides as the diameter, the inscribed $\angle ABC$ is 90°. In addition, because the triangle also has a 60° angle, its third angle must be 30°. Therefore, this is a 30°:60°:90° triangle with a hypotenuse of 40.

AB and BC are the legs of the right 30°:60°:90° triangle $\triangle ABC$, and can be determined using the properties of a 30°:60°:90° triangle, which include the side ratios: $x : x\sqrt{3} : 2x$, where:

  Hypotenuse $= z = 2x$

  Short leg $= x = z/2$

  Long leg $= y = x\sqrt{3} = (1/2)z\sqrt{3}$

We can use these properties along with $AC = 40$, to determine circle chords (sides of $\triangle ABC$) AB and BC:

Short leg $AB = 40/2 = 20$

Long leg $= BC = x\sqrt{3} = (1/2)z\sqrt{3} = 20\sqrt{3}$

Therefore, circle chord $AB = 20$ and circle chord $BC = 20\sqrt{3}$.

We can check this result using the Pythagorean Theorem

$\text{leg}^2 + \text{leg}^2 = \text{hypotenuse}^2$:

$20^2 + (20\sqrt{3})^2 = 40^2$, or $400 + 1{,}200 = 1{,}600$, or $1{,}600 = 1{,}600$

Triplet Right Triangles

• Other noteworthy right triangles include ***triplet right triangles***, also called ***Pythagorean triplets***, which are named according to their side lengths and satisfy the Pythagorean Theorem. The most common triplets are 3:4:5, 5:12:13, 7:24:25, and 8:15:17.

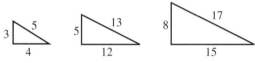

Multiples of the side ratios of these triangles are also triplets.

For example:

3:4:5	5:12:13	8:15:17	7:24:25
6:8:10	10:24:26	16:30:34	14:48:50
9:12:15			
12:16:20			
15:20:25			

4.9 Triangles and Trigonometric Functions

• Trigonometry is a visual and application-oriented field of mathematics that was developed by early astronomers and scientists to understand, model, measure, and navigate the physical world around them. Today, trigonometry has applications in numerous fields, including mathematics, astronomy, engineering, physics, chemistry, geography, navigation, surveying, architecture, and the study of electricity, light, sound, and phenomena with periodic and wave properties.

• Trigonometry involves measurements of angles, distances, triangles, arc lengths, circles, planes, spheres, and phenomena that exhibit a periodic nature. The six trigonometric functions, sine, cosine, tangent, cotangent, secant, and cosecant, can be defined using three different approaches: (1) As ratios of the sides of a right triangle, which is described in *Master Math: Trigonometry* Chapter 3; (2) In a coordinate system using angles in standard position, which is described in *Master Math: Trigonometry* Chapter 4; and (3) As arc lengths to a point on a unit circle called circular functions, which is described in *Master Math: Trigonometry* Chapter 4. Trigonometric functions also have a periodic nature that can be depicted on a graph, which is described in *Master Math: Trigonometry* Chapter 5. Trigonometric functions are determined, described, and illustrated in numerous venues including graphs, equations, vectors, polar coordinates, complex numbers, exponential functions, series expansions, and spherical surfaces.

• This section will introduce trigonometry and define the trigonometric functions as the ratios of the sides of a right triangle. You will also be introduced to the use of trigonometric functions in solving problems that involve setting up a model involving a right triangle in order to determine the length of a runway, distance to a star, the distance across a canyon, the angle of elevation of the Sun, and the distance of a ship to a lighthouse.

Right Triangle Definitions of the Trigonometric Functions

• Why should you care about the six trigonometric functions defined as ratios of the sides of a right triangle, especially sine, cosine, and tangent? Because they will help you find unknown distances and angles represented as the side lengths and angle measurements of triangles. There are selected applications included in this section with examples of using triangles to model problems. Determining the measures of sides and angles in a triangle is referred to as *solving the triangle*.

• *Right triangle relationships*: The **six trigonometric functions** (*sine, cosine, tangent, cotangent, secant,* and *cosecant*) can be defined according to the ratios of the three sides of a right triangle. A right triangle can be drawn alone or at the origin of a coordinate system. Consider the right triangle with sides x, y, and r:

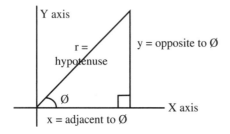

In this right triangle, angle Ø extends from the X-axis to the hypotenuse. (Ø is described in radians, or equivalently in degrees, and can be converted between the two.) If *r is the hypotenuse* and the terminal side of the angle Ø, *y is the side opposite Ø*, and *x is the side adjacent to Ø* and the initial side of angle Ø, then the *six trigonometric functions* are:

sine Ø = sin Ø = opposite/hypotenuse = y/r
cosine Ø = cos Ø = adjacent/hypotenuse = x/r
tangent Ø = tan Ø = opposite/adjacent = y/x = sin Ø / cos Ø
cosecant Ø = csc Ø = hypotenuse/opposite = r/y = 1 / sin Ø
secant Ø = sec Ø = hypotenuse/adjacent = r/x = 1/ cos Ø
cotangent Ø = cot Ø = adjacent/opposite = x/y = 1/ tan Ø

(To remember sin Ø = y/r, cos Ø = x/r, and tan Ø = y/x, think of the word SohCahToa or $S^o_h\,C^a_h\,T^o_a$ or $\mathbf{S}={}^{\mathbf{opp}}/_{\mathbf{hyp}}\ \mathbf{C}={}^{\mathbf{adj}}/_{\mathbf{hyp}}\ \mathbf{T}={}^{\mathbf{opp}}/_{\mathbf{adj}}.$) The *Pythagorean Theorem* written for this figure is $r^2 = x^2 + y^2$, and is used when solving right triangle problems involving trigonometric functions.

• In a right triangle, the trigonometric functions can be written with respect to *either of the acute angles*. Consider the following triangle with trigonometric functions for angle α rather than angle Ø.

sin α = opposite/hypotenuse = x/r
cos α = adjacent/hypotenuse = y/r
tan α = opposite/adjacent = x/y
csc α = hypotenuse/opposite = r/x
sec α = hypotenuse/adjacent = r/y
cot α = adjacent/opposite = y/x

Notice how these functions differ from the functions written for angle Ø listed above.

• *Sine, cosine, and tangent are most often used to find the angles and sides of right triangles and should be memorized.*

• *Definition*: The *cosine* of an acute angle of a right triangle is the ratio of the length of the *leg adjacent to the angle* to the length of the *hypotenuse*: cos Ø = adjacent/hypotenuse.

• *Definition*: The *sine* of an acute angle of a right triangle is the ratio of the length of the *leg opposite the angle* to the length of the *hypotenuse*: sin Ø = opposite/hypotenuse.

• *Definition*: The *tangent* of an acute angle of a right triangle is the ratio of the length of the *leg opposite the angle* to the length of the *leg adjacent to the angle*: tan Ø = opposite/adjacent.

• The *two acute angles in a right triangle* are always *complementary angles* (sum to 90°) because the angles in planar triangles sum to 180°.

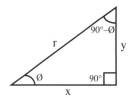

There are *cofunction identities* that *describe the complementary nature of the acute angles in a right triangle.* The trigonometric ratios of cosine, cotangent, and cosecant are the *cofunctions* of sine, tangent, and secant, respectively:

sin Ø = cos (90° – Ø)	cos Ø = sin (90° – Ø)
tan Ø = cot (90° – Ø)	cot Ø = tan (90° – Ø)
sec Ø = csc (90° – Ø)	csc Ø = sec (90° – Ø)

• There are also *reciprocal relationships* of the trigonometric functions that are helpful, especially when secant, cosecant, or cotangent are known and a calculation must be performed using a calculator that only has keys for sine, cosine, or tangent. The reciprocal relations of the trigonometric functions are:

csc Ø = 1 / sin Ø	or	sin Ø = 1 / csc Ø
sec Ø = 1 / cos Ø	or	cos Ø = 1 / sec Ø
cot Ø = 1 / tan Ø	or	tan Ø = 1 / cot Ø

These can be verified by multiplying both sides of each equation by the denominator and substituting the definitions of each function:

csc Ø sin Ø = 1 = (r/y)(y/r) = 1
sec Ø cos Ø = 1 = (r/x)(x/r) = 1
cot Ø tan Ø = 1 = (x/y)(y/x) = 1

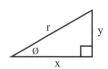

Solving Right Triangles

• *Solving right triangles* involves finding the measurements of the unknown sides and angles. The tools we can use to solve right triangles includes the trigonometric ratios, the fact that the two acute angles are complementary (sum to 90°), the reciprocal relationships given above, and the Pythagorean Theorem. There are usually several approaches that can be used to solve a right triangle, such as solving sides and angles in different orders or using different trigonometric functions.

• In general, when solving a right triangle consider the following:

1. In a right triangle, we know that the measure of one of the angles is always 90°.
2. If you know one of the acute angles, you can find the other angle because acute angles of a right triangle are complementary and sum to 90° (and the sum of all angles in a planar triangle is always 180°).
3. To find the first acute angle, use one of the trigonometric functions for sin Ø, cos Ø, tan Ø, csc Ø, sec Ø, or cot Ø, depending on which sides are known.
4. To find a side, use the trigonometric functions if an angle and a side are known.
5. To find the third side, use the Pythagorean Theorem or one of the six trigonometric functions in combination with a known angle.

For example, if we know the measures of one acute angle and one side, then we can find the other acute angle because the acute angles sum to 90°. Then we can find the sides using the trigonometric functions sine, cosine, or tangent, and the Pythagorean Theorem.

• To *determine the value* of sin Ø, cos Ø, tan Ø, csc Ø, sec Ø, or cot Ø, trigonometric tables have been calculated and are available in most trigonometry textbooks and on the internet. Modern scientific **calculators** with *trigonometric function keys* are a convenient tool for calculating the trigonometric functions. When using a *calculator* to solve *trigonometric functions* for the value of the function or for an angle, use the indicated buttons on the face of your calculator. Also make sure the calculator is in **degree mode** for calculations involving degrees or in **radian mode** for calculations involving radian angle measurements. (In introductory geometry classes you will generally be working only in degrees, but note that 2π radians = 360 degrees.)

• To calculate the cosine, sine, or tangent of an angle using a calculator, enter the angle measure and press the cosine, sine, or tangent key. To calculate the secant, cosecant, or cotangent of an angle, enter the angle and press the secant, cosecant, or cotangent key *if it is present* on the calculator.

If the secant, cosecant, and cotangent keys are *not* present on the calculator, then calculate the values using the *reciprocal relationships*:

csc Ø = 1/sin Ø	or	sin Ø = 1/csc Ø
sec Ø = 1/cos Ø	or	cos Ø = 1/sec Ø
cot Ø = 1/tan Ø	or	tan Ø = 1/cot Ø

• **Example:** Calculate (a) cos 30°, (b) sec 30°, (c) sin 30°, and (d) csc 30°.

(a) cos 30° ≈ 0.866

(b) sec 30° = 1/cos 30° ≈ 1/0.866 ≈ 1.1547

(c) sin 30° = 0.5

(d) csc 30° = 1/sin 30° = 1/0.5 = 2

• To calculate the value of the angle Ø using a calculator if the cosine,
sine, or tangent of an angle is known, use the appropriate *inverse function
key* on your calculator. For example:

If y = cos Ø, then Ø = cos⁻¹y.

If y = sin Ø, then Ø = sin⁻¹y.

If y = tan Ø, then Ø = tan⁻¹y.

The inverse keys are usually labeled as $\cos^{-1}$, $\sin^{-1}$, and $\tan^{-1}$, but may
also be identified as arccos, arcsin, or arctan. For secant, cosecant, or
cotangent, if the inverse keys are not present, use the *reciprocal relationships* (see previous example) to obtain the equivalent values of sine,
cosine, and tangent in the calculations.

• **Example:** Find α if (a) sin α = 1, or (b) sec α = 2.

(a) If sin α = 1, then α = sin⁻¹(1) = 90°

Therefore, α = 90°.

(b) If sec α = 2, then sec α = 1/cos α = 2, or 1/cos α = 2

2 = 1/cos α

2 cos α = 1

cos α = 1/2

α = cos⁻¹(1/2) = 60°

Therefore, α = 60°.

Examples and Applications of Right Triangles

• There are many questions that can be answered by setting up a model
involving a right triangle. The first two examples involve solving for
unknown values in a right triangle. Following these are applications that
use right triangles to find unknown distances and angles.

• **Example:** In this right triangle, find x, y, and α using your calculator and the trigonometric functions sin Ø = opposite/hypotenuse, cos Ø = adjacent/hypotenuse, or tan Ø = opposite/adjacent, as required.

Because the *sum of the angles in a planar triangle is 180°* and a right triangle has one 90° angle, then: 40° + α = 90°. Therefore α = 50°.

To find x and y use trigonometric ratios involving the known and unknown values.

To find y, use sin 40° = y/r = y/100 ft, or sin 40° = y/100 ft
Solve for y:
Therefore, y = (100 ft)(sin 40°) = 64.3 ft

To find x, use cos 40° = x/r = x/100 ft, or cos 40° = x/100 ft
Solve for x:
Therefore, x = (100 ft)(cos 40°) = 76.6 ft

In summary, the solution of this right triangle is: Ø = 40°, α = 50°, x = 76.6 ft, y = 64.3 ft, and r = 100 ft.

To check the side calculations, use the Pythagorean Theorem: $x^2 + y^2 = r^2$. Is it true that $76.6^2 + 64.3^2 = 100^2$? Yes.

• **Example:** Find x, Ø, and α using your calculator and the trigonometric functions sin Ø = opposite/hypotenuse, cos Ø = adjacent/hypotenuse, or tan Ø = opposite/adjacent, as required.

To find α, use sin α = y/r = 31.7 m / 60.2 m = 0.527, or sin α = 0.527
Therefore, α = $\sin^{-1}(0.527)$ = 31.8°.

To find Ø, use 90° − α = Ø
Therefore, Ø = 90° − 31.8° = 58.2°.

To find x, use tan α = y/x, or tan 31.8° = 31.7 m / x

Multiply both sides by x and divide both sides by tan 31.8°:

Therefore, x = (31.7 m) / (tan 31.8°) = 51.1 m.

In summary, the solution of this right triangle is: $\emptyset$ = 58.2°, α = 31.8°, x = 51.1 m, y = 31.7 m, and r = 60.2 m.

To check the side calculations, use the Pythagorean Theorem: $x^2 + y^2 = r^2$. Is it true that $51.1^2 + 31.7^2 = 60.2^2$? Yes, with rounding errors.

Note that we could have calculated x, given r and y, using the Pythagorean Theorem.

• **Example:** You need to find the distance *BC* across a canyon. Your friend is a geologist and you ask her for help. How does she find the distance?

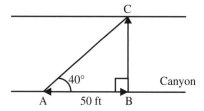

A geologist uses a surveyor's transit, or theodolite, which is an instrument having telescopic sight that is used to measure horizontal and vertical angles.

To measure the distance across a canyon, the geologist first measures a known length of 50 feet along one side of a canyon from point A to point B. She uses the transit at point B and sets a 90° angle to a point C. Then she uses the transit at point A and measures a 40° angle from B to C with its vertex at A.

Knowing two angles and a side of $\triangle$ABC, she determines the width of the canyon (length *BC*) using tan A = opposite / adjacent:

tan 40° = *BC* / 50 ft

Multiplying both sides by 50 ft results in the distance *BC*:

BC = (50 ft)(tan 40°) $\approx$ 42 feet across the canyon.

(Compare this example with the volcano-chasm-measuring example in Section 4.5 *Congruent Triangles*.)

• **Example:** You are building a runway that will head toward an existing 200-foot tower. If a jet takes off from the ground and ascends steadily at an angle of 20°, will it clear a 200 foot tower that is 400 feet from the point where the jet leaves the ground? If not, how far must the point of takeoff be from the tower?

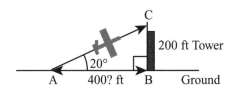

We assume ∠B is a right angle and are given angle A and the long leg of right triangle △ABC, so to determine whether the jet will clear the 200 foot tower, we need to calculate the altitude of the jet at the tower, or the length of *BC*. Given that the side adjacent to ∠A is 400 feet, we can determine the side opposite the ∠A (altitude) using the trigonometric function:

tan Ø = opposite / adjacent, or
tan 20° = *BC*/400

Multiply both sides by 400:

BC = (400 ft)(tan 20°) ≈ 146 feet

The jet will *not* clear the 200 foot tower!

For the jet to reach an altitude of 200 feet at *BC* of △ABC, we need to determine the distance *AB* that is required given the 20° takeoff angle. Again, we can use tan Ø = opposite / adjacent = tan 20° = 200 ft / *AB*

Multiply both sides by *AB* and divide both sides by tan 20°:

AB = 200 ft / tan 20° ≈ 550 feet

Therefore, the point of takeoff of the jet must be greater than 550 feet from the tower.

If we want a 100 foot safety clearance, the jet must have an altitude of 300 feet at the tower. Therefore, *BC* of △ABC is 300 so we need to determine *AB*. Again, we can use tan Ø = opposite / adjacent = tan 20° = 300 ft / *AB*

AB = 300 ft / tan 20° ≈ 824 feet

Therefore, to have a 100 foot safety clearance for the jet over the tower the point of takeoff of the jet must be 824 feet from the tower.

• **Example:** How can we measure the *distance to a star?* The most direct measurements of distance for stars are made with the *trigonometric parallax.* The trigonometric parallax is the apparent angular displacement of an object's position when viewed from two different locations. This technique works because when an object is viewed from two different locations, it appears to shift position. (Think about looking at an object through binoculars and closing one eye and then the other.) As the Earth moves in its orbit around the Sun, we observe stars from different points of view, and they appear to shift back and forth by differing amounts, depending on how far they are from us. Objects farther away exhibit a smaller shift. The parallax method measures the apparent displacement or dislocation of a star relative to a background field of much more distant objects out in space.

If we know the distance between the two locations where the observations are made (location of Earth in opposite sides of its orbit around the Sun), and if we can measure the angle through which a star appears to shift position (the angle of parallax) against background stars, then by using trigonometry, the distance from the Earth to the star can be calculated. Unfortunately, for stars that are very far away, the parallax method breaks down because the angle of parallax is so small. Other methods that rely on relative distances are used to estimate distances of stars, galaxies, etc.

To use the parallax method to determine a star's distance from Earth, the star's location is observed (using a telescope) against background stars at two opposite points in the Earth's orbit around the Sun. The annual parallax is defined as half of the angular shift (or angle of parallax) in a star's position against background stars when the Earth is at its two opposite locations in its orbit around the Sun, and is depicted by θ.

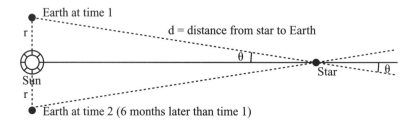

The distance from the star to the Earth, d, is calculated using the right triangle formed by Earth, the star, and the Sun, with r being the distance from Earth to Sun. To find d, we can use the simple *trigonometric relationship for a right triangle*:

$\sin θ = $ opposite / hypotenuse $ = $ r/d
$\sin θ = $ r/d
$d = $ r/$\sin θ$

In the early 1800s Friedrich Bessel found that the star 61 Cygni in the constellation Cygnus had a parallax of 0.314 arc seconds.

If we know that the distance from Earth to the Sun, r, is:

r = 150,000,000 km, and if we use Bessel's measurement of the parallax of star 61 Cygni in the constellation Cygnus as:

θ = 0.314", (where " denotes seconds), then we can calculate d, the distance from Earth to star 61 Cygni.

Using the conversion:

1 second = 1/60 of a minute = 1/3600 of a degree:
0.314 seconds / 3600 seconds per degree = 0.00008722°.

Therefore, the *distance from the Earth to the star* is:

d = r/sin θ = 150,000,000 km / sin(0.00008722°) = 9.85 × 10^{13} km.

We can convert this distance to *light years* by dividing kilometers by the speed of light multiplied by the number of seconds in a year:

9.85 × 10^{13} km/[(3 × 10^5 km/s)(3,600 s/hr × 24 hr/day × 365 day/yr)]

resulting in the distance to the star being approximately:

10 light years, which was the value Bessel calculated.

More recent measurements of the parallax of 61 Cygni resulted in 0.294" and a distance of 11.1 light years.

• Problems involving **angle of elevation** and **angle of depression** can be solved using a right triangle model. The angle of elevation is an angle measured from the *horizontal* upward, and an angle of depression is an angle measured from the *horizontal* downward.

• **Example** of *angle of elevation*: What is the approximate angle of elevation of the Sun, if a 100 foot telescope dome casts a shadow 70 feet long?

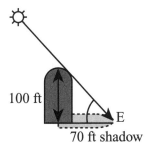

To find the angle of elevation, E, use the known sides of the formed right triangle and the tangent trigonometric ratio:

tan E° = opposite/adjacent = 100 ft / 70 ft
Therefore, the angle of elevation, E° = tan^{-1}(100/70) = 55°

• **Example** of *angle of depression*: You are in a lighthouse communicating with a ship offshore. You want to determine how far the ship is from the cliff that the lighthouse is on, given that the top of the lighthouse is 200 feet above sea level and the angle of depression is 20°. How would you do this?

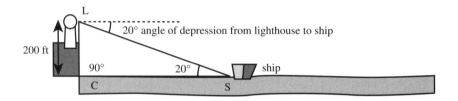

The angle of depression from the lighthouse to the ship is 20°; therefore, angle S is also 20° because they are alternate interior angles of parallel lines.

To find the distance from the cliff, C, to the ship, S, use the trigonometric ratio for a right triangle:

tan S° = opposite / adjacent = 200 ft / *CS*
tan 20° = 200 ft / *CS*
CS = 200 ft / tan 20° ≈ 550 feet

Therefore, the distance of the ship from the cliff and lighthouse is approximately 550 feet.

Oblique Triangles and the Law of Sines and Law of Cosines

• *Oblique triangles* are planar triangles that do not have a 90° angle, and therefore, are not right triangles. Oblique triangles may have all acute angles (<90°) or two acute angles and one obtuse angle (>90°). Like right triangles, oblique triangles can be used to model problems that require measurements of distances, lengths, and angles, such as determining the distance across a lake or canyon.

• Solving oblique triangles is helpful in situations where measuring distances and angles is required. To solve an oblique triangle and find all six measurements, the *Law of Sines*, the *Law of Cosines*, and the fact that the angles in a triangle sum to 180° are used. For the oblique triangle:

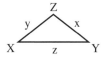

X, Y, and Z are the angles.
x, y, and z are the side lengths.

The sum of the angles is: $X + Y + Z = 180°$

The Law of Sines is: $(x/\sin X) = (y/\sin Y) = (z/\sin Z)$

The Law of Cosines is: $x^2 = y^2 + z^2 - 2yz \cos X$,

$y^2 = z^2 + x^2 - 2zx \cos Y$, and

$z^2 = x^2 + y^2 - 2xy \cos Z$

To find the angles X, Y, and Z, use the inverse cosine key $\cos^{-1}$ on your calculator. The laws of sine and cosine are derived in ***Master Math: Trigonometry***. The Law of Sines and the Law of Cosines can be used to solve right triangles, as well as oblique triangles, although the methods previously used to solve right triangles are easier to use when confronted with a right triangle. The Law of Sines and the Law of Cosines are derived from the principles of right triangle relations. Solving oblique triangles using the Law of Cosines and the Law of Sines is described in ***Master Math: Trigonometry*** Section 3.5.

4.10 Area of a Triangle

- The ***area of a triangle*** is written as:

$$\text{Area}_{(\text{triangle})} = (1/2)(\text{base})(\text{height})$$

$$\text{Area}_{(\text{triangle})} = (1/2)(\text{base})(\text{altitude})$$

The ***base*** is the ***side*** to which the *altitude* (or *height*) is drawn.

The ***height*** is the length of an ***altitude line segment*** that is drawn from a vertex angle and extends perpendicular to the side (or base) opposite that angle. The altitude (or height) is always perpendicular to its base.

To obtain the altitude (or height), draw a perpendicular line from the base to the opposite angle. In an oblique triangle, the altitude can be drawn perpendicular to the extension of the base.

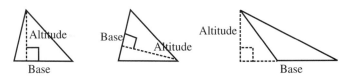

• *Triangle area theorem*: *The area of a triangle equals one-half the product of a side (base) and the altitude to that side.*

Area = (1/2) (length of a base) (length of the altitude to that base)

• **Example:** What is the area of the triangle?

Height/altitude = 3 in

Base = 6 in

(1/2)(6 inches)(3 inches) = 9 inches² or 9 square-inches, where the height is a perpendicular line from the vertex angle to the opposite base.

• *Corollary to triangle area theorem*: Triangles with congruent bases and congruent altitudes have equivalent areas.

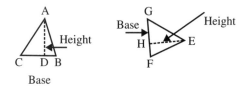

Area ΔABC = (*BC*)(*AD*)   Area ΔEFG = (*FG*)(*EH*)

If *BC* ≅ *FG* and *AD* ≅ *EH*, then Area ΔABC ≅ Area ΔEFG

• *Right triangle area theorem*: The area of a right triangle equals one-half the product of the length of its legs.

Area of right triangle = (1/2)(length of leg 1)(length of leg 2)

Leg 1   Hypotenuse
   Leg 2

Remember from Section 4.3. *Parts of Triangles*, that in a *right triangle*, two of the *altitudes* coincide with one of the legs, and area can be determined using either case in which an altitude coincides with one of the legs.

Area = (1/2)(altitude)(base)

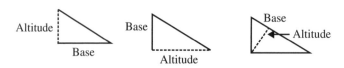

• *Side note*: The area of a parallelogram is: Area$_{\text{(parallelogram)}}$ = (base)(height)

The area of each triangle that makes up the parallelogram is:

Area$_{(triangle)}$ = (1/2)bh:

• *Median triangle area theorem*: A median divides a triangle into two triangles having equal areas.

(Remember, a **median** of a triangle is a line segment extended from any of the three angle vertexes to the *center or midpoint* of the side opposite the angle. A median *bisects* the side to which it is drawn.)

• *Common base triangle area theorem*: Triangles have equal areas if they share a common base, and their third vertices lie on a line that is parallel to their base.

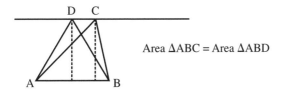

Because area = (1/2)(base)(altitude) and both triangles, ΔABC and ΔABD, have the same base and the same altitude length, they have the same area. (See Construction 28 in Chapter 8 *Constructions and Loci*.)

• **Example:** Find the area of the following equilateral triangle.

The altitude of an equilateral triangle divides it into two 30:60:90 triangles; in this case, with a hypotenuse of 4 and a short leg of 2. The long leg of each 30:60:90 triangle is the altitude of the equilateral triangle. We can find the long leg (or altitude) using either the Pythagorean Theorem, $x^2 + y^2 = z^2$, or the side proportions of a 30:60:90 triangle, which are: $x : x\sqrt{3} : 2x$.

Using Pythagorean Theorem: $2^2 + h^2 = 4^2$, $h^2 = 4^2 - 2^2 = 12$, $h = 2\sqrt{3}$

Using side proportions $x : x\sqrt{3} : 2x$, where $x = 2$: $h = 2\sqrt{3}$

Therefore, area $= (1/2)bh = (1/2)(4)(2\sqrt{3}) = 4\sqrt{3}$.

- In its simplest form, the *area of a triangle* is:

Area of a triangle $= (1/2)(base)(height)$, or $(1/2)(base)(altitude)$

However, if the altitude is *not* known, there are several other area formulas that are extensions of the above equation and can be employed to determine the ***area of a triangle***, depending on what information is known. Various formulas used to calculate area are described in ***Master Math*: *Trigonometry*** Section 3.7.

For example, if *three sides of a triangle are known*, ***Heron's formula*** can be used to determine the area of the triangle. Heron's formula is:

Area $= \sqrt{s(s-a)(s-b)(s-c)}$

where s is one-half of the perimeter of the triangle, called the "*semiperimeter*" and is given by: $s = (1/2)(a + b + c)$.

See ***Master Math*: *Trigonometry*** Section 3.7 for a derivation.

Heron's Theorem: The area of a triangle with side lengths a, b, and c, and semiperimeter $s = (1/2)(a + b + c)$ is:

Area $= \sqrt{s(s-a)(s-b)(s-c)}$

Example: Find the area of a triangle given three sides a $= 4$ meters, b $= 6$ meters, and c $= 8$ meters.

First, determine the semiperimeter s = (1/2)(a + b + c):

  s = (1/2)(4 + 6 + 8) = 9

Substitute into Heron's area formula, area = $\sqrt{s(s - a)(s - b)(s - c)}$:

  Area = $\sqrt{9(9 - 4)(9 - 6)(9 - 8)} = \sqrt{9(5)(3)(1)} = \sqrt{135}$

  Area ≈ 12 meters^2.

• We can derive a corollary from *Heron's Theorem* stating that the **area of an equilateral triangle** is: $(1/4)(\text{side}^2)\sqrt{3}$

To obtain this, let each side of an equilateral triangle be length a.

Substitute a into the semiperimeter and then into Heron's formula:

  s = (1/2)(a + b + c) = (1/2)(a + a + a) = 3a/2

  Area = $\sqrt{s(s - a)(s - b)(s - c)}$

  = $\sqrt{3a/2(3a/2 - a)(3a/2 - a)(3a/2 - a)}$

Where 3a/2 – a = 3a/2 – 2a/2 = a(3/2 – 2/2) = a/2

Substitute 3a/2 – a = a/2:

  Area = $\sqrt{3a/2(a/2)(a/2)(a/2)}$ = $(a/2)^2\sqrt{3}$ = $(1/4)a^2\sqrt{3}$

Therefore, the area of an equilateral triangle is: $(1/4)(\text{side})^2\sqrt{3}$

• The *area of a triangle* can be determined using several approaches depending on what information is known. For the triangle:

The simplest equation for area is: Area = (1/2)(base)(height), or area = (1/2)(b)(h), where h is the length of the altitude line perpendicular to the base that extends from the opposite vertex angle.

An extension of this equation, which calculates the height if it is unknown, can be used if two sides and the angle in between are known:

Area = (1/2)(b)(c sin A)

A further extension of this equation can be used if two angles and one side are known:

Area = (c^2 sin A sin B) / (2 sin C)

If three sides are known, Heron's formula can be used to find area:

Area = $\sqrt{s(s-a)(s-b)(s-c)}$, where s = (1/2)(a + b + c)

(See *Master Math: Trigonometry* Section 3.7 for derivations.)

4.11 Chapter 4 Summary and Highlights

• *Triangles* are three-sided polygons that contain three angles. Triangles can be classified as *right* or *oblique*. *Right triangles* are planar triangles with one angle equal to 90°, and *oblique triangles* are planar triangles that do *not* have a 90° angle. Types of triangles include *equilateral, isosceles, scalene, acute, obtuse,* and *right*.

• The *sum of the interior angles* in a planar triangle is always *180°*. The sum of the measures of the *exterior angles* of a triangle equals 360°. *Exterior/remote interior angle theorem*: The measure of an exterior angle of a triangle is equal to the *sum* of the measures of the two non-adjacent opposite (remote) interior angles.

• Triangles may possess the following: A *perpendicular bisector* *of a side*, which is perpendicular to and bisects that side at its *midpoint;* a *median*, which extends from an angle to the *midpoint* of the opposite side bisecting it; an *altitude*, which extends from an angle perpendicular to the opposite side; and an *angle bisector*, which bisects an angle.

• In a triangle, the largest side is opposite the largest angle, the smallest side is opposite the smallest angle, and the middle-length side is opposite the middle-size angle.

• *Congruent triangles* have the same size and same shape so that the three corresponding sides are of equal length, and the three corresponding angles have equal measure. There are certain postulates and theorems that can be used to determine whether two triangles are congruent. To determine *congruence* for two triangles, we need to identify at least three pairs of congruent parts, which must include at least one pair of known congruent side lengths. The three corresponding parts, which must be congruent, may be: three sides (SSS); two sides and their included angle (SAS);

two angles and their included side (ASA); or two angles and a non-included side (AAS). If the triangles are *right triangles*, the right angles are automatically congruent, so there are only two other pairs of corresponding parts that need to be congruent for the right triangles to be congruent. These include the hypotenuse and one leg (HL); the hypotenuse and one acute angle (HA); one leg and one acute angle (LA); or the two legs (LL).

• **Similar triangles** have their corresponding *angles congruent* and their corresponding *sides in proportion* (side ratios are equal). Similar triangles have the same shape, but not necessarily the same size.

• **AA similarity triangle postulate**: If *two* angles of one triangle are congruent to *two* angles of another triangle, then the triangles are *similar*. (The third angles will automatically be congruent.)

• Two triangles are similar if: All three pairs of corresponding angles are congruent; the corresponding sides have the same proportion; or one pair of corresponding angles is congruent and the corresponding sides including the angles are in proportion.

• **Triangle proportionality theorem**, also called the *side-splitter theorem*: If a line is parallel to one side of a triangle and intersects the other two sides, it divides the sides it intersects proportionally.

• **Midpoints parallel length theorem**: In a triangle, if a line connects the midpoints of two sides, then it will be parallel to the third side, and equal to half the length of the third side.

• If two similar triangles have the ratio of corresponding sides as x:y, then:

> the ratio of their perimeters is x:y;
>
> the ratio of any two corresponding segments (altitudes, medians, angle bisectors, etc.) is x:y; and
>
> the ratio of their areas is $x^2:y^2$.

• The *acute angles of a right triangle are* **complementary** and therefore sum to 90°.

• **Pythagorean Theorem**: In a right triangle, the square of the length of the hypotenuse is equal to the sum of the squares of the lengths of the legs: $(Leg)^2 + (Leg)^2 = (Hypotenuse)^2$

• Special *right triangles* include the **45:45:90 triangle** (having side ratios $x : x : x\sqrt{2}$), the **30:60:90 triangle** (having side ratios $x : x\sqrt{3} : 2x$), and the 3:4:5, 5:12:13, 7:24:25, and 8:15:17 Pythagorean triplet triangles.

• *Right triangle midpoint theorem*: In a right triangle, the midpoint of the hypotenuse is equidistant from the three vertices.

• The **six trigonometric functions** can be defined according to the *ratios of the sides of a right triangle*.

 y

cos α = adjacent/hypotenuse = x/r, sin α = opposite/hypotenuse = y/r,
tan α = opposite/adjacent = y/x, cot α = adjacent/opposite = x/y,
sec α = hypotenuse/adjacent = r/x, csc α = hypotenuse/opposite = r/y

• The **area of a triangle** can be determined using several approaches depending on what information is known. The simplest equation for area is: area = (1/2)(base)(height), or area = (1/2)(b)(h), where h is an altitude line extending from an angle perpendicular to the opposite base.

If three sides are known, *Heron's formula* can be used to find the area:

$$\text{area} = \sqrt{s(s-a)(s-b)(s-c)}\ , \text{ where } s = (1/2)(a+b+c).$$

Chapter

5

Polygons and Quadrilaterals

One of the most famous polygons is the ***golden rectangle***, which has a length and width that satisfy the equation:

$$\frac{l}{w} = \frac{l+w}{l} \quad \text{or} \quad \frac{length}{width} = \frac{length+width}{length}$$

The ratio *length*/*width* is the *golden ratio* (described in Chapter 3) and is equivalent to: $l/w = (1 + \sqrt{5})/2 \approx 1.618...$, which is the ***golden ratio*** Φ.

If a *golden rectangle* is divided into a square and a small rectangle, the outside and inside rectangles are *similar golden rectangles*.

$CD/CY = CY/YD = \Phi$
$AB/AX = AX/XB = \Phi$

Cutting a square from each successively smaller golden rectangle creates another smaller golden rectangle.

The **golden ratio** Φ is also associated with a *regular pentagon* and is equal to the ratio of the diagonal to the side.

$d/s = \Phi$

5.1 Polygons

• A *polygon* is a closed planar figure that is formed by three or more straight line segments that all meet at their endpoints. There are no endpoints that are not met by another endpoint in a polygon. The line segments that make up a polygon only intersect at their endpoints. Examples of polygons include:

Triangle,   Square,   Rectangle,   Octagon,   Hexagon,   Trapezoid,   Heptagon,   Pentagon

• *Polygons are named* according to the number of sides they contain.

3 sided polygon is a triangle.	8 sided polygon is an octagon.
4 sided polygon is a quadrilateral.	9 sided polygon is a nonagon.
5 sided polygon is a pentagon.	10 sided polygon is an decagon.
6 sided polygon is a hexagon.	12 sided polygon is a dodecagon.
7 sided polygon is a heptagon.	n sided polygon is a n-gon.

The term n-gon is often used to denote a polygon with n sides. Quadrilaterals include parallelograms, rectangles, rhombuses, squares, and trapezoids.

• In a polygon, the number of sides equal the number of angles.

• Polygons are either convex or concave (non-convex). **Convex polygons** have all angles being convex (pointing out) and **concave polygons** have one or more angles being concave (pointing in).

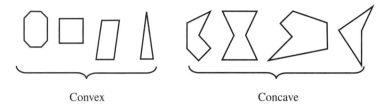

Convex   Concave

* **Parts of polygons**:

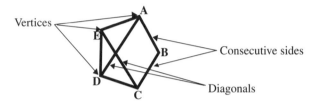

The polygon above, ABCDE, is a pentagon and contains the following:

Sides are the line segments that make up the polygon. In the figure, the sides are *AB*, *BC*, *CD*, *DE*, and *EA*.

Vertices are where the endpoints of the segments meet. A vertex, where two sides meet, forms an ***angle***. A *polygon is named* by listing its vertices consecutively either clockwise or counter-clockwise. In the figure, the vertices are A, B, C, D, and E.

Consecutive sides are any two sides (next to each other) that share an endpoint. In the figure, the consecutive sides are *AB* and *BC*, *BC* and *CD*, *CD* and *DE*, *DE* and *EA*, and *EA* and *AB*. ***Consecutive angles*** are any two angles that are next to each other in the polygon.

A ***diagonal*** of a polygon is a line segment that joins any two non-consecutive (not next to each other) vertices. The diagonals depicted in the figure are segments *AD* and *CE*.

• Certain features of triangles can be extended to other polygons. These include identifying or constructing the midpoint of a side, the perpendicular bisector of a side, an angle bisector, the medians, and the sum of the interior and exterior angles.

• One feature that pertains to triangles that does not apply to other polygons is that an equilateral triangle is always equiangular, and an equiangular triangle is always equilateral. This is not required for other polygons. For example, a quadrilateral may be only equilateral or only equiangular or both equilateral and equiangular:

| Equilateral and equiangular (square) | Equiangular but not equilateral (rectangle) | Equilateral but not equiangular (rhombus) |

5.2 Sum of the Interior and Exterior Angles in a Polygon

Interior Angles

• The equation for polygons that calculates the *sum of the interior angles* is:

$(n - 2)180°$ = Sum of all interior angles in n-gon

Where n is the number of angles (or sides) in the polygon.

Interior angle sum theorem: If a convex polygon has n sides, then the sum of its interior angles is given by the equation,

Interior angle sum $= (n - 2)180°$

• **Example**: A pentagon has five sides and five angles. What is the sum of all interior angle measurements?

The sum of all interior angle measurements is given by:

$(n - 2)180°$ = sum of all angles in n-gon

If n = 5, then: $(5 - 2)180° = (3)180° = 540°$

Therefore, the sum of the interior angles in a pentagon is 540°.

• There are $(n - 2)$ *non-overlapping triangles that can be drawn in a polygon*. For polygons having four or more sides, if all possible diagonals from one vertex are drawn, the polygon will be divided into non-overlapping triangles. The number of non-overlapping triangles will be *two* less than the number of sides. In other words, the number of triangles that can be drawn into a polygon of n sides equals $(n - 2)$ triangles.

We also know that the sum of the interior angles of any planer triangle is 180°, and it follows that the sum of the measures of the interior angles of a polygon equals the sum of the measures of the interior angles of the triangles it contains. This is consistent with our equation for the sum of the interior angles of a polygon:

Sum of interior angles of polygon = (n – 2)180°
Where n = number of sides in polygon.

A rectangle has	A pentagon has	A hexagon has
n – 2 triangles	n – 2 triangles	n – 2 triangles
= 4 – 2 = 2 triangles	= 5 – 2 = 3 triangles	= 6 – 2 = 4 triangles
(2)180° = 360°.	(3)180° = 540°.	(4)180° = 720°.
Angle sum =	Angle sum =	Angle sum =
(4 – 2)180° = 360°	(5 – 2)180° = 540°	(6 – 2)180° = 720°
(2)180° = 360°.	(3)180° = 540°.	(4)180° = 720°.

• **Example**: How many non-overlapping triangles does the following polygon contain, and what is the sum of its interior angle measurements?

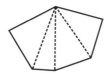

The polygon has 6 sides and is a hexagon.

There are (6 – 2) = 4 non-overlapping triangles. Therefore:

$$(4)180° = 720°$$

We can also use the sum of interior angle measurements given by:

$$(n – 2)180° = \text{sum of all interior angles in an n-gon}$$

If n = 6, then (6 – 2)180° = (4)180° = 720°.

Therefore, the hexagon contains 4 non-overlapping triangles, and the sum of its interior angles is 720°.

• **Example**: Determine the number of sides of a polygon if the sum of its interior angles is 3240°.

We can use the equation for the sum of all angles inside a polygon, (n – 2)180° = sum of all interior angles in n-gon, and solve for n:

$$(n – 2)180° = \text{sum of all interior angles}$$

Multiply 180° into parentheses and substitute interior angle sum:

$180°n - 360° = 3240°$
$180°n = 3600°$
$n = 3600°/180° = 20$

Therefore, the number of sides of a polygon having the sum of its interior angles as 3240° is 20 sides.

• **Example:** Determine the number of sides of a polygon if the sum of its interior angles is 1170°.

We can use the equation for the sum of all angles inside a polygon, $(n - 2)180°$ = sum of interior angles in n-gon, and solve for n:

$(n - 2)180°$ = sum of all interior angles

Multiply 180° into parentheses and substitute interior angle sum:

$180°n - 360° = 1170°$
$180°n = 1530°$
$n = 1530°/180° = 8.5$

The number of sides of a polygon cannot contain a 1/2 or 0.5 value, so there is no polygon having the sum of its interior angles as 1170°.

Exterior Angles

• An *exterior angle of a polygon* is an angle formed between the extension of one of the sides of the polygon and the outside of the polygon. Each exterior angle is a *supplement* of an adjacent interior angle so that an interior angle and its exterior angle sum to 180°:

Interior angle + Exterior angle = 180°

As a result, an exterior angle forms a *linear pair* (180° straight angle) with one of the interior angles.

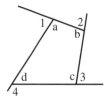

$L1$, $L2$, $L3$, and $L4$ are exterior angles.
La, Lb, Lc, and Ld are interior angles.
$mLa + mL1 = 180°$, $mLb + mL2 = 180°$, $mLc + mL3 = 180°$, and $mLd + mL4 = 180°$, form linear pairs.

• *Exterior angles polygon sum theorem*: **The sum of the measures of the exterior angles, one at each vertex, of any convex polygon, equals 360°.** (This is true for a convex polygon having any number of sides.)

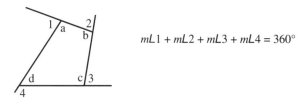

$$mL1 + mL2 + mL3 + mL4 = 360°$$

• Note that if $L1$, $L2$, $L3$, and $L4$ above are aligned side by side with one vertex, they will form a circle, and circles always measure 360° around.

• **Example:** Determine the sum of the exterior angles of the following convex polygons: (a) heptagon, (b) nonagon, (c) triangle, (d) quadrilateral, and (e) 26-gon.

(a) 360°, (b) 360°, (c) 360°, (d) 360°, and (e) 360°.

• Because each exterior angle is a supplement of an adjacent interior angle, if you calculate the supplements of the interior angles and sum them, you can prove to yourself that the exterior angles sum to 360°. For example, if the interior angles are 140°, 50°, 130°, and 40°, then their supplements (exterior angles) are 40°, 130°, 50°, and 140°, respectively.

$$140° + 50° + 130° + 40° = 360° \quad \text{Interior angles in the quadrilateral}$$

$$\underline{40° + 130° + 50° + 140° = 360°} \quad \text{Exterior angles in the quadrilateral}$$

$$180° \qquad 180° \qquad 180° \qquad 180°$$

5.3 Regular Polygons and Their Interior and Exterior Angle Measures

• Snowflakes are natural manifestations of geometry, and are often shaped as regular hexagons.

• A regular (equilateral) triangle, a regular quadrilateral (square), and a regular hexagon can be placed together in a continuous pattern or *tiling*, which covers a surface without any intervening spaces.

• A ***regular polygon*** is both equilateral and equiangular. In other words, if the lengths of the sides are equal to each other, and the angle measurements are equal to each other, the polygon is a *regular polygon*. For example, a *square* having four congruent sides and four congruent angles is a *regular quadrilateral*. A *pentagon* having five congruent sides and five congruent angles is a *regular pentagon*. Regular polygons are always *convex*.

Each figure has congruent angles and congruent sides

• ***Definition***: A ***regular polygon*** is a convex polygon that is both equilateral and equiangular.

• Regular polygons include:

3 equal sides & angles: Equilateral triangle.	8 equal sides & angles: Regular octagon.
4 equal sides & angles: Square.	9 equal sides & angles: Regular nonagon.
5 equal sides & angles: Regular pentagon.	10 equal sides & angles: Regular decagon.
6 equal sides & angles: Regular hexagon.	12 equal sides & angles: Regular dodecagon.
7 equal sides & angles: Regular heptagon.	n equal sides & angles: Regular n-gon.

Definitions and Properties of Regular Polygons

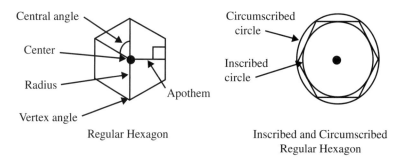

Regular Hexagon   Inscribed and Circumscribed
Regular Hexagon

• **Definition**: The **center** is a point on the interior of a regular polygon that is equidistant from its vertices. The *center* of a regular polygon is the common center of its circumscribed circle and its inscribed circle.

• **Definition**: The **radius** of a regular polygon is a line segment that joins its center to any vertex. The radius of a regular polygon is also the radius of a circumscribed circle. The *radii* of a regular polygon are *congruent*. A radius of a regular polygon bisects the **vertex angle** to which it is drawn. (Each angle on the polygon is a vertex angle, and, in a regular polygon, they are congruent to each other.)

• **Definition**: The **apothem** of a regular polygon is a line segment extending from its center perpendicular to one of its sides. The *apothem* can extend from the center to any side and is a perpendicular bisector of that side. The apothem is also the radius of an inscribed circle. The *apothems* of a regular polygon are *congruent*. An *apothem* of a regular polygon *bisects* the side to which it is drawn. The apothem is used to calculate the area of a regular polygon. Area$_{(regular\ n\text{-}gon)}$ = (1/2)(apothem)(perimeter). (See Section 5.8. *Area and Perimeter*.)

• *Side note*: An apothem and a radius of certain regular polygons form characteristic 45:45:90 and 30:60:90 right triangles that can be used to determine the lengths of the radius and apothem of a polygon.

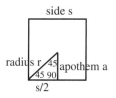

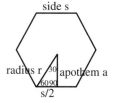

45:45:90 triangle with side ratios:

$x : x : x\sqrt{2} = s/2 : s/2 : (s/2)\sqrt{2}$

Therefore: $a = s/2$ and $r = (s/2)\sqrt{2}$

30:60:90 triangle with side ratios:

$x : x\sqrt{3} : 2x = s/2 : (s/2)\sqrt{3} : 2(s/2)$

Therefore: $a = (s/2)\sqrt{3}$ and $r = 2(s/2) = s$

• *Definition*: The ***central angle*** of a ***regular polygon*** is an angle formed by two radii drawn to two consecutive vertices.

For a regular polygon with n congruent sides, the measure of each central angle equals 360°/n.

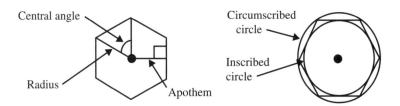

• ***Regular cyclic polygon theorem***: **Every regular polygon is cyclic.**

If a circle is drawn through the vertices of a regular polygon, the distance from the center to each vertex will be the same.

A circle can be *circumscribed* around any regular polygon.

A circle can be *inscribed* inside any regular polygon.

The *center of a circle* circumscribed about a regular polygon is also the center of a circle inscribed in that same polygon.

• If an ***equilateral polygon*** can be inscribed inside a circle, then the equilateral polygon is also a regular polygon.

• The ***angles in a regular polygon*** having n sides obey the following equations, which are discussed in the succeeding pages in subsection *Regular Polygons: Interior and Exterior Angle Measures*.

Each interior angle $i_{\text{regular n-gon}}$ = 180°(n − 2)/n

Each exterior angle $e_{\text{regular n-gon}}$ = 360°/n

Interior angle i + exterior angle e = 180°

Each central angle $c_{\text{regular n-gon}}$ = 360°/n

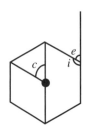

Regular Polygons: Interior and Exterior Angle Measures

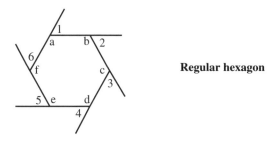

Regular hexagon

In a *regular polygon* all interior angle measures are equal to each other:

$$La \cong Lb \cong Lc \cong Ld \cong Le \cong Lf$$

In a *regular polygon* all exterior angle measures are equal to each other:

$$L1 \cong L2 \cong L3 \cong L4 \cong L5 \cong L6$$

Interior Angle Measure for Regular Polygons

• Because all interior angle measures are equal in a regular polygon, to find the *measure of an interior angle in a regular polygon*, we can determine the *interior angle sum* and divide it by the number of sides.

Interior angle $i_{\text{regular n-gon}}$ = **180°(n – 2)/n**

Where n is the number of sides in the regular polygon and (n – 2)180° is the sum of the interior angles in any convex polygon.

• **Example:** Determine the measure of Lc in the regular hexagon using the *interior angle sum*.

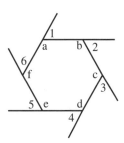

To determine the measure of one of the interior angles, we can calculate the *interior angle sum* and divide it by the number of sides, which is 6 for a hexagon.

To calculate the sum use the equation for the sum of all interior angle measurements: $(n - 2)180° =$ sum of all angles in n-gon

For a hexagon n $= 6$ therefore, $(6 - 2)180° = (4)180° = 720°$.

The sum of the angles in any convex hexagon is 720°, and because this is a regular hexagon having congruent angles, any one of the angle measures is $720° \div 6 = 120°$.

These steps are equivalent to using the equation for a regular polygon:

Interior angle $i_{\text{regular n-gon}} = 180°(n - 2)/n = 180°(6 - 2)/6 = 120°$

Therefore, the measure of $\angle c$ in the a regular hexagon is 120°.

• An alternative means to finding the ***measure of an interior angle in a regular polygon*** is to use the *exterior angle sum* (which is always 360°). We first calculate the *measure of an exterior angle* by dividing 360° by the number of sides. Then, because the interior and exterior angles are supplements (sum to 180°), each interior angle is equal to:

Interior angle measure $= 180° -$ (exterior angle measure)

• **Example:** Find the measure of $\angle c$ in the regular hexagon using the *exterior angle sum.*

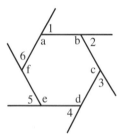

To determine the measure of one of the interior angles in the regular hexagon, we can first use the *exterior angle sum*, 360°, to obtain the *measure of an exterior angle* by dividing 360° by the number of sides, which is 6: $360° \div 6 = 60°$.

Therefore, 60° is the *measure of an exterior angle in the regular hexagon.*

Then, because the interior and exterior angles are supplements (sum to 180°), each interior angle is equal to:

Interior angle measure $= 180° -$ exterior angle measure, or
$180° - 60° = 120°$

Therefore, the measure of $\angle c$ in the regular hexagon is 120°, which we obtained in the preceding example.

Exterior Angle Measure For Regular Polygons

• *In a regular polygon all interior angle measures are equal to each other*, which means *all exterior angle measures are also equal to each other*. Because the *sum* of the *exterior angles in a convex polygon is always 360°*, the **measure of each exterior angle** can be determined by dividing 360° by the number of sides.

$$\textbf{Exterior angle } e_{\textbf{regular n-gon}} = \textbf{360°/n}$$

Where n is the number of sides in the polygon.

In addition:

$$\textbf{Interior angle } i + \textbf{exterior angle } e = \textbf{180°}$$

This is true because each exterior angle is a *supplement* of an adjacent interior angle so that an interior angle and its exterior angle form a straight line and sum to 180°.

• **Example:** Find the measure of $L5$ in the regular hexagon.

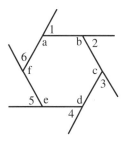

To determine the measure of one of the exterior angles in the regular hexagon, we can divide the exterior angle sum, 360°, by the number of sides, which is 6: 360° ÷ 6 = 60°

Therefore, 60° is the *measure of an exterior angle in the regular hexagon.*

Remember the equation of an exterior angle of a regular polygon is:
Exterior angle $e_{\text{regular n-gon}}$ = 360°/n

Therefore, the measure of $L5$ in the regular hexagon is 60°.

• **Example:** For a regular polygon with 12 sides, a dodecagon, what is the measure of each interior angle i and each exterior angle e?

Interior angle $i_{\text{regular n-gon}}$ = 180°(n – 2)/n = 180°(12 – 2)/12 = 150°

Exterior angle $e_{\text{regular n-gon}}$ = 360°/n = 360°/12 = 30°

Therefore, interior angle i = 150° and exterior angle e = 30°.

Also, does: interior angle i + exterior angle $e = 180°$?

   $150° + 30° = 180°$. Yes.

• **Example:** How many sides does a regular polygon have if one of its exterior angles measures $10°$?

We can use the exterior angle formula for a regular polygon to determine n:
Exterior angle $e_{\text{regular n-gon}} = 360°/n$

   Substitute $e = 10°$
   $10° = 360°/n$

Solve for n by multiplying both sides by n and then dividing by $10°$:

   $n = 360°/10° = 36$

Therefore, a regular polygon having an exterior angle measure of $10°$ has 36 sides.

5.4 Quadrilaterals

• *Definition*: A quadrilateral is a polygon that has four sides.

• *Quadrilaterals* are four-sided polygons that can have different shapes and sizes. Quadrilaterals include parallelograms, rhombuses, rectangles, squares, and trapezoids. The following figure depicts the relationship between them:

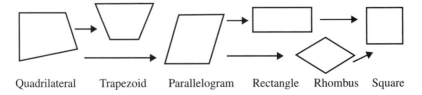

Quadrilateral   Trapezoid   Parallelogram   Rectangle   Rhombus   Square

A *quadrilateral* has four sides.

A *trapezoid* has exactly one pair of parallel bases or sides.

A *parallelogram* has two pairs of parallel sides and congruent opposite angles.

A *rectangle*, which is a parallelogram, has both pairs of opposite sides parallel and four congruent right angles. (Equiangular.)

A *rhombus*, which is a parallelogram, has all four sides of equal length and opposite angles congruent. (Equilateral.)

A *square*, which is a parallelogram, has all four sides of equal length and four congruent right angles. (Equiangular and equilateral.)

• *Quadrilateral interior angle sum theorem*: The sum of the measures of the interior angles in all quadrilaterals is 360°.

We can demonstrate this using the equation for the sum of the interior angles in a polygon: $(n - 2)180° = (4 - 2)180° = (2)180° = 360°$.

• **Example:** Prove that the sum of the interior angles in a quadrilateral is 360°.

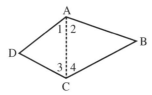

Given: Quadrilateral ABCD.

Prove: $m\angle A + m\angle B + m\angle C + m\angle D = 360°$.

Plan: Draw diagonal *AC* to create two triangles. $m\angle 1 + m\angle 3 + m\angle D = 180°$ and $m\angle 2 + m\angle 4 + m\angle B = 180°$. Combine the angles of the two triangles which sum to 360° and substitute $m\angle A = m\angle 1 + m\angle 2$ and $m\angle C = m\angle 3 + m\angle 4$ resulting in $m\angle A + m\angle B + m\angle C + m\angle D = 360°$.

Proof:

Statements	Reasons
1. Quadrilateral ABCD.	1. Given.
2. Draw diagonal *AC*.	2. Two points determine a line.
3. $m\angle 1 + m\angle 3 + m\angle D = 180°$; $m\angle 2 + m\angle 4 + m\angle B = 180°$.	3. The sum of angles in any Δ is 180°.
4. $m\angle 1 + m\angle 3 + m\angle D + m\angle 2 + m\angle 4 + m\angle B = 360°$.	4. Addition of equations in #3.
5. $m\angle A = m\angle 1 + m\angle 2$; $m\angle C = m\angle 3 + m\angle 4$.	5. Angle addition postulate.
6. $m\angle A + m\angle B + m\angle C + m\angle D = 360°$.	6. Substitution of #4 and #5.

5.5 Parallelograms

• A *parallelogram* is a quadrilateral in which both pairs of opposite sides are parallel to each other. Each pair of opposite sides is called *bases of the parallelogram*. The symbol for a parallelogram is: ▱

• The *quadrilateral interior angles sum theorem* applies to parallelograms:
The sum of measures of the angles in a quadrilateral is 360°.

Definitions and Theorems for Parallelograms

• *Definition*: The opposite sides of a parallelogram are parallel.

If *AB* ‖ *CD* and *AD* ‖ *BC*,
then ABCD is a parallelogram.

• *Parallelogram congruent sides theorem*: The opposite sides of a
parallelogram are congruent.

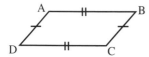

If *AB* ≅ *CD* and *AD* ≅ *BC*,
then ABCD is a parallelogram.

Prove the *parallelogram congruent sides theorem*:

Given: Parallelogram ▱ABCD.

Prove: *AB* ≅ *CD* and *AD* ≅ *BC*.

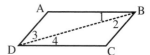

Proof:

Statements	Reasons
1. ▱ABCD.	1. Given.
2. Draw diagonal *BD*.	2. Two points determine a line.
3. *AB* ‖ *CD* and *AD* ‖ *BC*.	3. Definition of parallelogram.
4. ∠1 ≅ ∠4 and ∠2 ≅ ∠3.	4. Parallel lines cut by transversal (*BD*) create ≅ alternate interior angles.
5. *BD* ≅ *BD*.	5. Identity or reflexive axiom: Self = self.
6. ΔABD ≅ ΔCDB.	6. ASA Δ congruence post: If 2 ∠s and the included side of one Δ are ≅ to corresponding parts of another Δ, Δs are ≅.
7. *AB* ≅ *CD* and *AD* ≅ *BC*.	7. Corresponding parts of ≅ Δs are ≅.

• *Parallelogram congruent angles theorem*: The *opposite angles of a
parallelogram* are congruent.

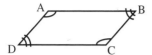

If $LA \cong LC$ and $LB \cong LD$,
then ABCD is a parallelogram.

Prove the *parallelogram congruent angles theorem*:

Given: Parallelogram $\square$ABCD.

Prove: $LA \cong LC$ and $LB \cong LD$.

Plan: Because opposite sides are parallel, consecutive interior angles are supplementary. It follows that $LA \cong LC$ and $LB \cong LD$ because LA and LC are both supplementary to LD, and LB and LD are both supplementary to LC, and angles supplementary to the same angle are congruent.

Proof:

Statements	Reasons
1. $\square$ABCD.	1. Given.
2. $AB \parallel CD$.	2. Definition of parallelogram.
3. LA is supplementary to LD; LB is supplementary to LC.	3. Parallel lines (#2) cut by transversal (*AD* & *BC*) create supplementary consecutive interior Ls.
4. $AD \parallel BC$.	4. Definition of parallelogram.
5. LD is supplementary to LC; LA is supplementary to LB.	5. Parallel lines (#4) cut by transversal (*AB* & *CD*) create supplementary consecutive interior Ls.
6. $LA \cong LC$ and $LB \cong LD$.	6. Supplements to same angle are $\cong$. LA & LC suppl to LD; LB & LD suppl to LC.

• *Parallelogram supplementary angles theorem*: Any two *consecutive angles of a parallelogram* are supplementary.

If $mLA + mLB = 180°$, $mLB + mLC = 180°$, $mLC + mLD = 180°$, and $mLA + mLD = 180°$, then ABCD is a parallelogram.

The parallelogram supplementary angles theorem is true because opposite sides of a parallelogram are, by definition, parallel, and the consecutive interior angles of each pair of opposite sides are supplementary. The principle behind this is that parallel lines (either pair of opposite sides) cut by a transversal (other pair of opposite sides) create supplementary consecutive interior angles. Each pair of opposite sides are the parallel lines and the other pair of sides act as transversals. (See *parallel transversal postulate* discussion in Chapter 2 Section 2.3. *Parallel Lines*.)

• ***Parallelogram diagonals bisect theorem***: The *diagonals of a parallelogram bisect* each other.

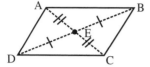

If *AE* ≅ *EC* and *DE* ≅ *EB*,
then ABCD is a parallelogram.

Prove *parallelogram diagonals bisect theorem*:

Given: Parallelogram □ABCD.

Prove: Diagonals *AC* and *DB* bisect each other.

Plan: Show that ΔAEB ≅ ΔCED leading to *AE* ≅ *EC* and *DE* ≅ *EB*.
Then, by definition of a bisector, *AC* and *DB* bisect each other.

Proof:

Statements	Reasons
1. □ABCD.	1. Given.
2. *AB* ∥ *CD*.	2. Definition of parallelogram.
3. ∟BAC ≅ ∟DCA and ∟ABD ≅ ∟CDB.	3. Parallel lines (#2) cut by transversal (*AC* & *DB*) create congruent alternate interior angles.
4. *AB* ≅ *CD*.	4. Opposite sides of a □ are congruent.
5. ΔAEB ≅ ΔCED.	5. ASA Δ congruence post: If 2 ∟s and the included side of one Δ are ≅ to corresponding parts of another Δ, Δs are ≅.
6. *AE* ≅ *EC* and *DE* ≅ *EB*.	6. Corresponding parts of ≅ Δs are ≅.
7. *AC* and *DB* bisect each other.	7. Def. of bisector: Cuts into 2 ≅ segments.

• ***Parallelogram congruent triangles theorem***: A *diagonal of a parallelogram* divides it into two congruent triangles.

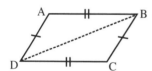

If ΔABD ≅ ΔCDB
then ABCD is a parallelogram.

• Summary of definitions and theorems for parallelograms:

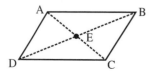

In $\square$ABCD:

AB is parallel to *CD*. *AD* is parallel to *BC*.

Opposite sides are congruent: $AB \cong CD$ and $AD \cong BC$.

Opposite angles are congruent: $\angle A \cong \angle C$ and $\angle B \cong \angle D$.

Consecutive interior angles are supplementary: $m\angle A + m\angle B = 180°$, $m\angle B + m\angle C = 180°$, $m\angle C + m\angle D = 180°$, and $m\angle A + m\angle D = 180°$.

Diagonals bisect each other: $AE \cong EC$ and $DE \cong EB$.

A diagonal divides into two congruent triangles: $\triangle ABD \cong \triangle CDB$ or $\triangle BAC \cong \triangle DCA$.

• An ***altitude of a parallelogram*** is a line segment that is perpendicular to both of its bases. The length of the altitude is called the ***height of the parallelogram***. The altitude line may be perpendicular to an extension of one of the bases.

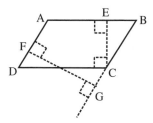

Altitude *EC* is perpendicular to bases *AB* and *CD*.

Altitude *EC* is the height of $\square$ABCD with *AB* and *CD* as bases.

Altitude *FG* is perpendicular to base *AD* and the extension of *BC*.

Altitude *FG* is the height of $\square$ABCD with *AD* and *BC* as bases.

• **Example:** In $\square$ABCD, if $m\angle A = 120°$, $AB = 6$, and $AD = 4$, what is $\angle B, \angle C, \angle D, BC,$ and *CD*?

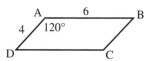

To solve this problem we can use the following three theorems:

    The opposite sides of a parallelogram are congruent.

    Any two consecutive angles of a parallelogram are supplementary.

    The opposite angles of a parallelogram are congruent.

To determine *BC* and *CD* we can use the fact that opposite sides are congruent:

    $AB = CD = 6$ and $AD = BC = 4$

Because consecutive angles are supplementary, we can determine LB using:

$mLA + mLB = 180°$, or $120° + mLB = 180°$

Solving for LB:

$mLB = 180° - 120° = 60°$

Because opposite angles are congruent, we can determine LD using $LB \cong LD$ where $mLB = 60°$; therefore, $mLD = 60°$.

Because opposite angles are congruent, $LA \cong LC$ where $mLA = 120°$; therefore, $mLC = 120°$.

Therefore, $mLB = 60°$, $mLC = 120°$, $mLD = 60°$, $BC = 4$, and $CD = 6$.

Proving Quadrilaterals Are Parallelograms

• There are several ways to prove that a quadrilateral is a parallelogram. They are first summarized and then explained in theorems and their proofs.

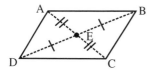

In summary, a quadrilateral is a parallelogram ($\square$ABCD) if:

1. Both pairs of opposite sides are parallel:

   $AB \parallel CD$ and $AD \parallel BC$.

2. Both pairs of opposite sides are congruent:

   $AB \cong CD$ and $AD \cong BC$.

3. Both pairs of opposite angles are congruent:

   $LA \cong LC$ and $LB \cong LD$.

4. One pair of opposite sides is parallel and congruent:

   $AB \parallel CD$ and $AB \cong CD$ or $AD \parallel BC$ and $AD \cong BC$.

5. Its diagonals bisect each other: $AE \cong EC$ and $DE \cong EB$.

6. Consecutive interior angles are supplementary:

   $mLA + mLB = 180°$,
   $mLB + mLC = 180°$,
   $mLC + mLD = 180°$, and
   $mLA + mLD = 180°$.

• By *definition*, *if opposite sides of a quadrilateral are parallel, it is a parallelogram.*

This definition and the following theorems provide the means to prove that a quadrilateral is a parallelogram.

• *Parallelogram sides theorem*: A quadrilateral is a parallelogram if both pairs of opposite *sides* are congruent.

Prove *parallelogram sides theorem*:

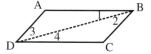

Given: ABCD: $AB \cong CD$ and $AD \cong BC$.

Prove: Quadrilateral ABCD is a parallelogram.

Proof:

Statements	Reasons
1. ABCD: $AB \cong CD$ and $AD \cong BC$. 2. Draw diagonal *BD*. 3. $BD \cong BD$. 4. $\triangle ABD \cong \triangle CDB$. 5. $L1 \cong L4$ and $L2 \cong L3$. 6. $AB \parallel CD$ and $AD \parallel BC$. 7. ABCD is a parallelogram.	1. Given. 2. Two points determine a line. 3. Identity or reflexive axiom: Self = self. 4. SSS $\triangle$ congruence post: If 3 sides of one $\triangle$ are $\cong$ to corresponding parts of another $\triangle$, the $\triangle$s are $\cong$. 5. Corresponding parts of $\cong$ $\triangle$s are $\cong$. 6. If a transversal between two lines forms $\cong$ alternate interior Ls, the lines are parallel. 7. A quadrilateral is a parallelogram if both pairs of opposite sides are $\parallel$.

• *Parallelogram angles theorem*: A quadrilateral is a parallelogram if both pairs of opposite *angles* are congruent.

Prove *parallelogram angles theorem*:

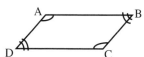

Given: ABCD: $LA \cong LC$ and $LB \cong LD$.

Prove: Quadrilateral ABCD is a parallelogram.

Proof:

Statements	Reasons
1. ABCD: $\angle A \cong \angle C$ and $\angle B \cong \angle D$.	1. Given.
2. $m\angle A + m\angle C + m\angle B + m\angle D = 360°$.	2. Sum of angles in quadrilateral is 360°.
3. $m\angle A + m\angle A + m\angle B + m\angle B = 360°$.	3. Substitution of #1 and #2.
4. $2m\angle A + 2m\angle B = 360°$, or $m\angle A + m\angle B = 180°$.	4. Rearrange #3 using algebra.
5. $\angle A$ and $\angle B$ are supplementary.	5. Supplementary angles sum to 180°.
6. $AD \parallel BC$.	6. If 2 lines cut by transversal form consecutive supplementary interior $\angle$s, lines are $\parallel$.
7. $2m\angle A + 2m\angle D = 360°$, or $m\angle A + m\angle D = 180°$.	7. Substitute and rearrange #1 & #2.
8. $\angle A$ and $\angle D$ are supplementary.	8. Definition of supplementary $\angle$s.
9. $AB \parallel CD$.	9. If 2 lines cut by transversal form consecutive supplementary interior $\angle$s, lines are $\parallel$.
10. ABCD is a parallelogram.	10. A quadrilateral is a parallelogram if both pairs of opposite sides are parallel. #6, #9.

• *Parallelogram 2 sides theorem*: A quadrilateral is a parallelogram if one pair of opposite sides is both congruent and parallel.

Prove *parallelogram 2 sides theorem*:

Given: ABCD: $AB \cong CD$ and $AB \parallel CD$.

Prove: Quadrilateral ABCD is a parallelogram.

Proof:

Statements	Reasons
1. ABCD: $AB \cong CD$ & $AB \parallel CD$.	1. Given.
2. Draw diagonal BD.	2. Two points determine a line.
3. $BD \cong BD$.	3. Identity or reflexive axiom: Self = self.
4. $\angle 1 \cong \angle 2$.	4. If two $\parallel$ lines are cut by a transversal (BD), alternate interior $\angle$s are congruent.
5. $\triangle ABD \cong \triangle CDB$.	5. SAS $\triangle$ congruence post: If 2 sides and the included $\angle$ of one $\triangle$ are $\cong$ to corresponding parts of another $\triangle$, the $\triangle$s are $\cong$. (#1,#3,#4)
6. $AD \cong CB$.	6. Corresponding parts of $\cong$ $\triangle$s are $\cong$.
7. ABCD is a parallelogram.	7. If opposite sides of quadrilateral are $\cong$, it is a parallelogram.

• *Parallelogram diagonals theorem*: A quadrilateral is a parallelogram if its *diagonals bisect* each other.

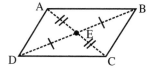

If $AE \cong EC$ and $DE \cong EB$,
then ABCD is a parallelogram.

Prove *parallelogram diagonals theorem*:

Given: Diagonals AC and DB bisect each other.

Prove: ABCD is a parallelogram.

Proof:

Statements	Reasons
1. AC and DB bisect each other.	1. Given.
2. $AE \cong EC$ and $DE \cong EB$.	2. Def of bisector: Cuts into 2 $\cong$ segments.
3. $\angle AEB \cong \angle CED$; $\angle AED \cong \angle CEB$.	3. Vertical angles are congruent.
4. $\triangle AEB \cong \triangle CED$; $\triangle AED \cong \triangle CEB$.	4. SAS $\triangle$ congruence post: If 2 sides and the included $\angle$ of one $\triangle$ are $\cong$ to corresponding parts of another $\triangle$, the $\triangle$s are $\cong$.
5. $AB \cong CD$ and $AD \cong CB$.	5. Corresponding parts of $\cong$ $\triangle$s are $\cong$.
6. ABCD is a parallelogram.	6. If opposite sides of quadrilateral are $\cong$, it is a parallelogram.

• *Parallelogram supplementary theorem*: A quadrilateral is a parallelogram if all four pairs of consecutive interior angles are supplementary.

If $m\angle A + m\angle B = 180°$, $m\angle B + m\angle C = 180°$,
$m\angle C + m\angle D = 180°$, and $m\angle A + m\angle D = 180°$,
then ABCD is a parallelogram.

The *parallelogram supplementary theorem* is true because parallel lines cut by a transversal create supplementary consecutive interior angles. Similarly, if consecutive supplementary angles are formed by two lines cut by a transversal, then the lines are parallel. Therefore, if consecutive interior angles are supplementary in a quadrilateral, then the lines (sides) cut by a transversal (other sides) are parallel. Then, because if opposite sides of a quadrilateral are parallel, it is a parallelogram.

5.6 Special Parallelograms: Rectangles, Rhombuses, and Squares

• Rectangles, rhombuses, and squares are all parallelograms and have the following descriptions:

A *rectangle*, *rhombus*, or *square* has all the properties of a parallelogram.

A *rectangle* is an equiangular parallelogram with four right angles.

A *rhombus* is an equilateral parallelogram with four congruent sides.

A *square* is an equiangular and equilateral parallelogram with four right angles and four congruent sides.

A *square* is a rectangle, a rhombus, and a parallelogram.

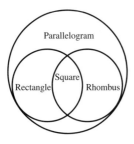

Table of diagonal properties of rectangles, rhombuses, and squares

Diagonal Properties	Parallelogram	Rectangle	Rhombus	Square
Diagonals bisect each other.	√	√	√	√
A diagonal forms 2 congruent triangles.	√	√	√	√
Diagonals are congruent. (Equal lengths.)		√		√
Diagonals are perpendicular.			√	√
Diagonals bisect vertex angles.			√	√
Diagonals form 4 congruent triangles.			√	√

Rectangles

• A *rectangle* is a quadrilateral with all right angles.

• The minimum basic definition of a *rectangle* is that:

A rectangle is a parallelogram with one right angle.

This is true because *consecutive angles of a parallelogram are supplementary* and if one angle is a right angle, then the remaining angles must also be right angles.

If ABCD is a parallelogram and $m\angle A = 90°$, then ABCD is a rectangle.

• ***Quadrilateral interior angles sum theorem***: The sum of measures of the interior angles of a quadrilateral is 360°.

A ***corollary to quadrilateral interior angles sum theorem*** is that: Each angle of a rectangle is a right angle. (360° ÷ 4 = 90°)

Therefore, a rectangle is an equiangular parallelogram with all four angles having equal measurements of 90° so that *each angle of a rectangle is a right angle*.

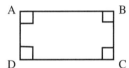

$m\angle A = m\angle B = m\angle C = m\angle D = 90°$

• ***Rectangles are parallelograms theorem***: All rectangles are parallelograms.

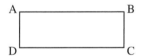

Informally prove *rectangles are parallelograms theorem*:

Given: ABCD is a rectangle.

Prove: ABCD is a parallelogram.

Proof:

   $m\angle A = 90°$, $m\angle B = 90°$, and $m\angle C = 90°$ because a rectangle is a quadrilateral with all right angles.

   $m\angle A + m\angle B = 180°$. Therefore, $\angle A$ and $\angle B$ are supplementary.

   $m\angle C + m\angle B = 180°$. Therefore, $\angle C$ and $\angle B$ are supplementary.

   $\angle A$ and $\angle B$ are consecutive interior angles of AD and BC. Two lines are parallel if when cut by a transversal, supplementary angles are formed. Therefore, $AD \parallel BC$.

LB and LC are consecutive interior angles of AB and CD. Two lines are parallel if when cut by a transversal, supplementary angles are formed. Therefore, $AB \parallel CD$.

Because $AD \parallel BC$ and $AB \parallel CD$, then using the definition of a parallelogram, that opposite sides are parallel, ABCD is a parallelogram.

• Because *rectangles are* **parallelograms**, they possess the properties of a parallelogram, which include:

The *opposite sides of a parallelogram* (rectangle) are parallel.

The *opposite sides of a parallelogram* (rectangle) are congruent.

The *opposite angles of a parallelogram* (rectangle) are congruent.

Any two *consecutive angles of a parallelogram* (rectangle) are supplementary.

The *diagonals of a parallelogram* (rectangle) bisect each other.

A *diagonal of a parallelogram* (rectangle) divides it into two congruent triangles.

• **Diagonals rectangle theorem**: The *diagonals of a rectangle* are congruent.

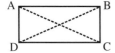

$$AC \cong DB$$

If ABCD is a parallelogram and $AC \cong DB$, then ABCD is a rectangle (or a square).

Prove *diagonals rectangle theorem*:

Given: ABCD is a rectangle.

Prove: Diagonals AC and DB are congruent.

Proof:

Statements	Reasons
1. ABCD is a rectangle.	1. Given.
2. $AB \cong CD$.	2. Opposite sides of a rectangle are $\cong$.
3. $AD \cong DA$.	3. Identity or reflexive axiom: Self = self.
4. LBAD and LCDA are Rt angles.	4. All angles in a rectangle are right angles.
5. LBAD $\cong L$CDA.	5. All right angles are congruent.
6. ΔBAD $\cong \Delta$CDA.	6. SAS Δ congruence post: If 2 sides and the included L of one Δ are $\cong$ to corresponding parts of another Δ, the Δs are $\cong$.
7. $AC \cong DB$.	7. Corresponding parts of $\cong \Delta$s are $\cong$.

- The *length of the diagonals of a rectangle* can be determined using the Pythagorean Theorem if the side lengths of the rectangle are known.

  (long-side length)^2 + (short-side length)^2 = (diagonal length)^2

- **Example:** In rectangle ABCD, if *AC* = 20 yards, find *DB*, *AE*, and *DE*.

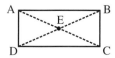

In a rectangle, diagonals are congruent; therefore, $AC \cong DB$. Because $AC \cong DB$, if *AC* = 20 yards, then *DB* = 20 yards.

Because a rectangle is a parallelogram, and the diagonals of a parallelogram bisect each other, then $AE \cong EC$ and $DE \cong EB$. It follows that:

AE = (1/2)AC = (1/2)20 yards = 10 yards
EC = (1/2)AC = (1/2)20 yards = 10 yards
DE = (1/2)DB = (1/2)20 yards = 10 yards
EB = (1/2)DB = (1/2)20 yards = 10 yards

Therefore, *DB* = 20 yards, *AE* = 10 yards, and *DE* = 10 yards.

Rhombus

- A *rhombus* is a quadrilateral with congruent sides. In other words, all the sides of a *rhombus* are of equal length. A rhombus is also an equilateral *parallelogram*. A *rhombus* is a *parallelogram* with all four sides of equal length. In addition, opposite *angles* of a rhombus are equal to each other (which is true for all parallelograms).

- The minimum basic definition of a *rhombus* is that:

A rhombus is a parallelogram in which two adjacent sides are congruent.

- ***Rhombus adjacent sides theorem***: If two adjacent sides of a parallelogram are congruent, the parallelogram is a rhombus.

If ABCD is a parallelogram and $AB \cong BC$, then ABCD is a rhombus.

This is true because opposite sides of a parallelogram are congruent. Therefore, if two adjacent sides of a parallelogram are congruent, then all four sides must be congruent and the parallelogram is a rhombus. In addition, if adjacent sides are congruent and opposite sides are parallel, then all four sides must be congruent.

• **Rhombuses are parallelograms theorem**: All *rhombuses* are parallelograms.

Informally prove *rhombuses are parallelograms theorem*:

Given: ABCD is a rhombus.

Prove: ABCD is a parallelogram.

Proof: $AB \cong CD$ and $AD \cong BC$ because a rhombus is a quadrilateral with congruent sides.

ABCD is a parallelogram because a quadrilateral that has opposite sides congruent, is a parallelogram.

• Because all *rhombuses are parallelograms*, they posses the properties of a parallelogram, which include:

   The *opposite sides of a parallelogram* (rhombus) are parallel.

   The *opposite sides of a parallelogram* (rhombus) are congruent.

   The *opposite angles of a parallelogram* (rhombus) are congruent.

   Any two *consecutive angles of a parallelogram* (rhombus) are supplementary.

   The *diagonals of a parallelogram* (rhombus) bisect each other.

   A *diagonal of a parallelogram* (rhombus) divides it into two congruent triangles.

• The *quadrilateral interior angles sum theorem* applies to rhombuses: The sum of measures of the angles of a quadrilateral (rhombus) is 360°.

• Additional properties of a rhombus: Its **diagonals are perpendicular** to each other and they also **bisect opposite angles**.

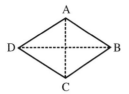

$AC \perp DB$;
AC bisects $\angle DAB$ and $\angle DCB$;
DB bisects $\angle ADC$ and $\angle ABC$.

• **_Rhombus diagonals theorem_**: The *diagonals* of a *rhombus* are perpendicular bisectors of each other.

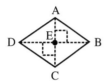

$AC \perp DB$; $AE \cong EC$; $DE \cong EB$.

• **Example:** Prove that if the diagonals of a parallelogram are perpendicular to each other, the parallelogram is a rhombus (or square).

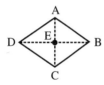

Given: ABCD is a parallelogram; $AC \perp DB$.

Prove: ABCD is a rhombus.

Proof:

Statements	Reasons
1. ABCD is a parallelogram.	1. Given.
2. AC and DB bisect each other.	2. The diagonals of a ▱ bisect each other.
3. $DE \cong EB$.	3. Definition of bisect: Cut in 2 ≅ segments.
4. $AE \cong AE$.	4. Identity or reflexive axiom: Self = self.
5. $AC \perp DB$.	5. Given.
6. $\angle AED$ and $\angle AEB$ are Rt angles.	6. Perpendicular lines form 4 right angles.
7. $\angle AED \cong \angle AEB$.	7. All right angles are congruent.
8. $\triangle AED \cong \triangle AEB$.	8. SAS Δ congruence post: If 2 sides and the included ∠ of one Δ are ≅ to corresponding parts of another Δ, the Δs are ≅. (#3, #4, #7)
9. $AD \cong AB$.	9. Corresponding parts of ≅ Δs are ≅.
10. ABCD is a rhombus.	10. A ▱ with 2 adjacent ≅ sides is a rhombus.

• ***Rhombus diagonal angles theorem***. The *diagonals* of a *rhombus* bisect the vertex angles.

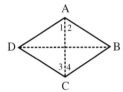

$AC \perp DB$;
AC bisects $\angle DAB$ and $\angle DCB$
and DB bisects $\angle ADC$ and $\angle ABC$.

Prove *rhombus diagonal angles theorem* for one diagonal:

Given: ABCD is a rhombus.

Prove: AC bisects $\angle DAB$ and $\angle DCB$.

Proof:

Statements	Reasons
1. ABCD is a rhombus.	1. Given.
2. $AD \cong DC$ and $AB \cong BC$.	2. A rhombus is a $\square$ with $\cong$ adjacent sides.
3. $\angle 1 \cong \angle 3$ and $\angle 2 \cong \angle 4$.	3. In a Δ (ΔADC, ΔABC), angles opposite 2 congruent sides are $\cong$ to each other.
4. $AB \parallel CD$ and $AD \parallel BC$.	4. Opposite sides of a $\square$ are parallel.
5. $\angle 1 \cong \angle 4$ and $\angle 2 \cong \angle 3$.	5. If two parallel lines are cut by a transversal, alternate interior $\angle$s are $\cong$.
6. $\angle 1 \cong \angle 2$ and $\angle 3 \cong \angle 4$.	6. Substitution of #3 and #5.
7. AC bisects $\angle DAB$ and $\angle DCB$.	7. Definition of $\angle$ bisector: Cuts into 2 $\cong$ $\angle$s. (#6 and $m\angle 1 + m\angle 2 = m\angle DAB$; $m\angle 3 + m\angle 4 = m\angle DCB$)

• **Example:** In rhombus ABCD, if $AB = 4$ and $\angle ABC = 80°$, find BC, $\angle DAB$, $\angle AEB$, and $\angle ABE$.

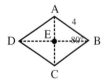

To solve this problem we can use the following:

All the sides of a rhombus are congruent.

The diagonals of a rhombus bisect the vertex angles.

The diagonals of a rhombus are perpendicular bisectors.

Any two consecutive angles of a parallelogram (or rhombus) are supplementary.

Because the sides of a rhombus are congruent: $AB = BC = 4$

Because the diagonals of a rhombus are perpendicular:

$\angle AEB$ is a right angle measuring 90°.

Because the diagonals of a rhombus bisect the vertex angles:

$m\angle ABE = (1/2)m\angle ABC = (1/2)80° = 40°$

Because any two consecutive angles are supplementary: If $m\angle ABC = 80°$ and $m\angle ABC + m\angle DAB = 180°$, then $m\angle DAB = 180° - 80° = 100°$.

Therefore, $BC = 4$, $m\angle DAB = 100°$, $m\angle AEB = 90°$, and $m\angle ABE = 40°$.

• **The *diagonals* of a *rhombus* form four congruent triangles.** Furthermore, because the diagonals of a rhombus are perpendicular bisectors of each other, these triangles are *right triangles*.

$$\triangle AED \cong \triangle AEB \cong \triangle CED \cong \triangle CEB$$

• **Example:** If the diagonals in the above rhombus measure $AC = 15$ m and $DB = 20$ m, what is the perimeter of rhombus ABCD?

Because the diagonals of a rhombus are perpendicular bisectors and form four congruent right triangles, we can calculate a side of rhombus ABCD using the Pythagorean Theorem for right triangles and then multiply by four to obtain perimeter.

For $\triangle AEB$, AB is the hypotenuse and using the Pythagorean Theorem: $AB^2 = AE^2 + BE^2$

If $AC = 15$, then $AE = 15/2 = 7.5$
If $DB = 20$, then $BE = 20/2 = 10$
$AB^2 = (7.5)^2 + (10)^2 = 56.25 + 100 = 156.25$
$AB \approx 12.5$ meters

If side AB is 12.5 meters, then the perimeter of rhombus ABCD is:

4(12.5 meters) = 50 meters.

Square

• A *square* is an equilateral and equiangular parallelogram. In other words, a square is a parallelogram in which the four angles and four sides are congruent. A square is also described as a rectangle in which two adjacent sides are congruent. A square is a *regular polygon*.

• A square has all the properties of both a *rhombus* and a *rectangle*. A square is a quadrilateral with four congruent 90° angles and four congruent sides, (which combines the properties of a rhombus and a rectangle). A square is both a rectangle and a rhombus.

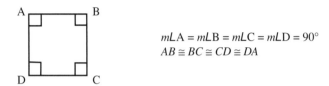

$$mLA = mLB = mLC = mLD = 90°$$
$$AB \cong BC \cong CD \cong DA$$

• The minimum basic definition of a **square** is that:

A square is a parallelogram with a right angle and two congruent adjacent sides.

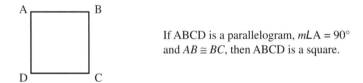

If ABCD is a parallelogram, $mLA = 90°$ and $AB \cong BC$, then ABCD is a square.

• Because a square is both a rectangle and a rhombus, as well as a parallelogram, a square possesses the properties of a parallelogram, which include:

The *opposite sides of a parallelogram* (square) are parallel.

The *opposite sides of a parallelogram* (square) are congruent.

The *opposite angles of a parallelogram* (square) are congruent.

Any two *consecutive angles of a parallelogram* (square) are supplementary.

The *diagonals of a parallelogram* (square) bisect each other.

A *diagonal of a parallelogram* (square) divides it into two congruent triangles.

• In addition, the *quadrilateral interior angles sum theorem* applies to squares: The sum of measures of the interior angles of a quadrilateral is 360°.

• The **diagonals of a square** are congruent. Furthermore, because the angles in a square each measure 90°, the length of the diagonals can be determined using the *Pythagorean Theorem* for right triangles providing the side lengths of the square are known:

$$\textbf{(side length)}^2 + \textbf{(side length)}^2 = \textbf{(diagonal length)}^2$$

If the side length is x and the diagonal length is d:

$x^2 + x^2 = d^2$, or $d^2 = 2x^2$, or $d = \sqrt{2x^2} = x\sqrt{2}$

• All the *diagonal properties* for rectangles, rhombuses, and parallelograms apply to squares:

Diagonals bisect each other.

A diagonal forms 2 congruent triangles.

Diagonals are congruent.

Diagonals are perpendicular.

Diagonals bisect vertex angles.

Diagonals form 4 congruent triangles.

Proving a Parallelogram Is a Rectangle, Rhombus, or Square

• The minimum requirements for proving a parallelogram is a rectangle, rhombus, or a square are summarized as follows:

If a parallelogram has *one right angle*, then it is a ***rectangle***.

If a parallelogram has *congruent diagonals*, then it is a ***rectangle***.

If a parallelogram has *congruent adjacent sides*, then it is a ***rhombus***.

If a parallelogram has *a right angle and two congruent adjacent sides*, then it is a ***square***. (*A square is both a rectangle and rhombus.*)

5.7 Trapezoids

• Bridges and railroad trestles have been constructed using trapezoid-shaped geometry during the past two centuries.

• A ***trapezoid*** is a quadrilateral having one and *only* one pair of opposite sides parallel to each other. These parallel sides are called ***bases***. The other pair of opposite sides are not parallel and are called ***legs***. The angles that include each base are called *base angles*.

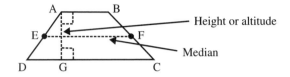

AB is parallel to *CD*.

AD is not parallel to *BC*.

AB and *CD* are called *bases*.

AD and *BC* are called *legs*.

AG is an *altitude* and is perpendicular to bases *AB* and *CD*.

E and F are midpoints of *AD* and *BC*, respectively.

EF is the *median* of trapezoid ABCD.

∠A and ∠B represent one pair of *base angles*.

∠D and ∠C represent one pair of *base angles*.

• An *altitude of a trapezoid* is a line segment that is perpendicular to both of the bases. The length of the altitude is called the *height of the trapezoid*.

• The *median of a trapezoid* is a line segment that joins the *midpoints* of the legs and is parallel to the bases.

Trapezoid median theorem: The median of any trapezoid has two properties:

(1) It is parallel to both bases.

(2) Its length equals half the sum of the base lengths:
    Median = (1/2)(Base 1 + Base 2)

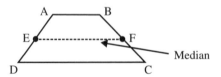

EF is the median of trapezoid ABCD.

E is the midpoint of *AD*; F is the midpoint of *BC*.

AB ∥ *CD*; *AB* ∥ *EF*; *EF* ∥ *CD*.

EF = (1/2)(*AB* + *CD*).

• **Example:** In trapezoid ABCD, if $AB = 5$ and $CD = 13$, what is median *EF*?

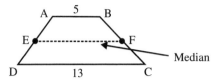

To find *EF* we can use the trapezoid median theorem: The length of the median of any trapezoid equals half the sum of the base lengths.

$$EF = (1/2)(AB + CD)$$
$$EF = (1/2)(5 + 13) = 9$$

Therefore, $EF = 9$.

Isosceles Trapezoid

• An *isosceles trapezoid* is a trapezoid in which the non-parallel sides (legs) have equal lengths.

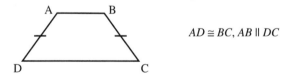

$AD \cong BC$, $AB \parallel DC$

• The *base angles of an isosceles trapezoid* are the angles at the ends of the bases.

Isosceles trapezoid base angle theorem: The *base angles* of an *isosceles trapezoid* are congruent.

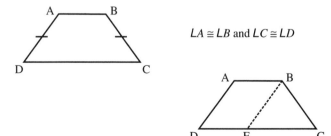

$LA \cong LB$ and $LC \cong LD$

Prove *isosceles trapezoid base angle theorem*:

Given: ABCD is an isosceles trapezoid with legs $AD \cong BC$.

Prove: Base angles: $\angle A \cong \angle B$ and $\angle C \cong \angle D$.

Plan: Draw $BE \parallel AD$ forming $\square ABED$ and isosceles $\triangle BEC$. Show $\angle BEC \cong \angle C$ and $\angle D \cong \angle BEC$ and therefore $\angle D \cong \angle C$. $\angle A$ and $\angle D$ are supplements and $\angle B$ and $\angle C$ are supplements; therefore, $\angle A \cong \angle B$ because supplements of congruent angles ($\angle D \cong \angle C$) are congruent.

Proof:

Statements	Reasons
1. ABCD is an isosceles trapezoid.	1. Given.
2. Draw $BE \parallel AD$ to form ABED.	2. Through a point not on a line there is exactly one line parallel to that line.
3. $AB \parallel CD$.	3. Def of trapezoid: bases are parallel.
4. ABED is a parallelogram.	4. Def of $\square$: Opposite sides parallel. #2, #3.
5. $AD \cong BE$.	5. Opposite sides of a $\square$ are congruent.
6. $AD \cong BC$.	6. Given.
7. $BC \cong BE$.	7. Trans. Prop. Substitution of #5, #6.
8. $\triangle BEC$ is an isosceles triangle.	8. Isosceles triangle has 2 $\cong$ sides.
9. $\angle BEC \cong \angle C$.	9. Base angles of isosceles $\triangle$ are $\cong$.
10. $\angle D \cong \angle BEC$.	10. If two $\parallel$ lines ($BE \parallel AD$) are cut by a transversal, corresponding $\angle$s are $\cong$.
11. $\angle D \cong \angle C$.	11. Trans. Prop. Substitution of #9 and #10.
12. $\angle A$ and $\angle D$ are supplements; $\angle B$ and $\angle C$ are supplements.	12. Parallel lines cut by transversal (legs) create supplementary consecutive interior $\angle$s.
13. $\angle A \cong \angle B$.	13. Supplements of $\cong$ $\angle$s ($\angle D$&$\angle C$) are $\cong$.

• **Converse of isosceles trapezoid base angle theorem**: If the base angles of a trapezoid are congruent, the *trapezoid* is *isosceles*.

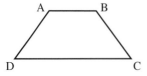

In isosceles trapezoid ABCD, if $m\angle D = m\angle C$ and $m\angle A = m\angle B$, then $AD \cong BC$.

• **Example:** In isosceles trapezoid ABCD, if the ratio of $\angle A$ to $\angle D$ is 3x to 2x, find the value of x, y, and z.

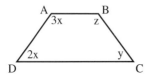

Because $AB \parallel DC$, then $m\angle A + m\angle D = 180°$. We can write an equation using the ratio of $\angle A$ to $\angle D$, $3x : 2x$, and $m\angle A + m\angle D = 180°$ to solve for x:

$3x + 2x = 180°$

$x(3 + 2) = 180°$

$x = 180°/5 = 36°$

Substitute $x = 36°$ into $3x$ and $2x$ to obtain $\angle A$ and $\angle D$:

$m\angle A = 3(36°) = 108°$

$m\angle D = 2(36°) = 72°$

Because base angles of an isosceles trapezoid are congruent:

$m\angle A = m\angle B = z = 3x = 108°$

$m\angle D = m\angle C = y = 2x = 72°$

Therefore, $x = 36°$, $y = 72°$, and $z = 108°$.

• **Isosceles trapezoid diagonals theorem**: The *diagonals* of an *isosceles trapezoid* are congruent.

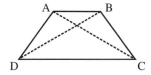

$AC \cong BD$

Prove *isosceles trapezoid diagonals theorem*:

Given: ABCD is an isosceles trapezoid with legs $AD \cong BC$.

Prove: Diagonals $AC \cong BD$.

Plan: Show that $\triangle ADC \cong \triangle BCD$ using SAS. Then $AC \cong BD$ because corresponding parts of congruent triangles are congruent.

Proof:

Statements	Reasons
1. ABCD is an isosceles trapezoid.	1. Given.
2. $AD \cong BC$.	2. Legs of isosceles trapezoid are $\cong$.
3. $\angle C \cong \angle D$.	3. Base $\angle$s of isosceles trapezoid are $\cong$.
4. $DC \cong CD$.	4. Identity or reflexive axiom: Self = self.
5. $\triangle ADC \cong \triangle BCD$.	5. SAS $\triangle$ congruence postulate: If 2 sides and the included $\angle$ of one $\triangle$ are $\cong$ to corresponding parts of other $\triangle$, $\triangle$s are $\cong$.
6. $AC \cong BD$.	6. Corresponding parts of $\cong$ $\triangle$s are $\cong$.

• **Example:** In trapezoid ABCD, if $AD \cong BC$, $m\angle DAB = 110°$ and $DB = 5000$, what is AC and $m\angle ABC$?

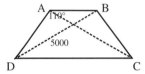

Because $AD \cong BC$, trapezoid ABCD is an isosceles trapezoid.

Because the diagonals of an isosceles trapezoid are congruent:

  $DB = AC = 5000$.

Because base angles of an isosceles trapezoid are congruent:

  $m\angle DAB = m\angle ABC = 110°$.

Therefore, $AC = 5000$ and $m\angle ABC = 110°$.

5.8 Area and Perimeter of Squares, Rectangles, Parallelograms, Rhombuses, Trapezoids, Other Polygons, and Regular Polygons

• The **perimeter** *of polygons and planar figures* is the sum of the lengths of its sides or the distance around. (For a circle, the perimeter is called *circumference*.) The units for perimeter represent length or distance, and are singular because of the one-dimension being measured or described. Remember to convert all measurements to the same units (such as feet, meters, inches, centimeters, miles, etc.) before adding, subtracting, multiplying, or dividing.

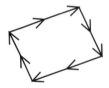

Perimeter is the distance around.

• The **area** *of polygons and other planar figures* is a measure of its internal size in the two dimensions of its planar surface. The units of area represent two-dimensional space and are squared, describing the square units of space contained within the figure's surface. Examples of square units include square feet (ft^2), square meters (m^2), square miles (mi^2), and special units such as acres.

1 unit

1 square unit | 1 unit

The area is the number of square units within the surface of a planar figure. Remember to convert all measurements to the same units, such as (feet)^2, (meters)^2, (inches)^2, (centimeters)^2, (miles)^2, etc., before adding, subtracting, multiplying, or dividing.

Area (shaded) is the two-dimensional space.

• Polygonal regions can be divided into smaller triangular regions.

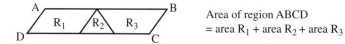

• *Area addition postulate*: The area of an entire region is the sum of the areas of non-overlapping regions within it.

Area of region ABCD
= area R_1 + area R_2 + area R_3

Area and Perimeter of a Square

• The *perimeter of a square* is the sum of its four sides, or equivalently, four times the length of one side.

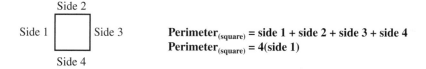

Side 2

Side 1 | | Side 3

Side 4

Perimeter$_{(square)}$ = side 1 + side 2 + side 3 + side 4
Perimeter$_{(square)}$ = 4(side 1)

• The *area of a square* is the length of a side squared or (side)^2.

Area square postulate: The area of a square is the square of a side length, or Area$_{(square)}$ $= s^2 =$ (side)(side)

Side s

| | Side s

Area$_{(square)}$ = (side)^2

• **Example:** What is the area of the square if each small square is 1 inch by 1 inch?

The area is $(side)^2 = (3 \text{ inches})^2 = 9 \text{ inches}^2 = 9$ square inches.

In this example, we can also count 9 square-inch units in the figure to visualize the area.

• The *area of a square* can also be given by: **Area = (1/2)(diagonal)²**

We can illustrate this by applying the formula for the length of the diagonals in a square (described at the end of our discussion on squares in Section 5.6): $d^2 = 2x^2$.

Substitute $d^2 = 2x^2$ into Area $= (1/2)(\text{diagonal})^2$:

Area $= (1/2)d^2 = (1/2)(2x^2) = x^2$, where x is the side length.

$$\textbf{Area}_{(\textbf{square})} = \textbf{(1/2)(d)}^2 = \textbf{(side)}^2$$

• **Example:** Find the area of a square if the radius of a circle circumscribing the square is 15.

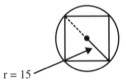

If the radius of the circle is 15, then the diameter is twice the radius:

d = 2r = (2)(15) = 30.

The diameter of the circle is equal in length to the diagonal of the square inside the circle. To find the area of the square, we can use the equation:

$$\text{Area}_{(\text{square})} = (1/2)(d)^2$$

$$\text{Area}_{(\text{square})} = (1/2)(30)^2 = 450$$

Therefore, the area of the square is 450.

• **Example:** If a warehouse building is a perfect square with a side length of 100 yards, what is the perimeter and area of the building?

100 yards

100 yards

$\text{Perimeter}_{(\text{square})} = 4(\text{side}) = 4(100 \text{ yards}) = 400 \text{ yards}.$

$\text{Area}_{(\text{square})} = \text{side}^2 = (100 \text{ yards})^2 = 10,000 \text{ yards}^2 \text{ or } 10,000 \text{ square yards}.$

• **Example:** If the perimeter of a square is 16 miles, what is its area?

Perimeter = 16 miles

The perimeter of a square is given by:

$\text{Perimeter}_{(\text{square})} = 4(\text{side length})$

By rearranging: Side length $= (\text{perimeter})/4$

If perimeter $= 16$ miles, then side length $= 16/4 = 4$ miles.

The area of a square is given by: $\text{Area} = \text{side}^2$

For a side length of 4 miles this becomes:

$\text{Area}_{(\text{square})} = (4 \text{ mi})^2 = 16 \text{ mi}^2 = 16 \text{ square miles}.$

Therefore, the area of a square with a 16 mile perimeter is 16 square miles.

Area and Perimeter of a Rectangle

• The ***perimeter of a rectangle*** is the sum of its four sides, or equivalently, two-times the long sides (l) plus two-times the short sides (h).

Side 2, l

Side 1, h Side 3, h

Side 4, l

$\text{Perimeter}_{(\text{rectangle})} = \text{side 1} + \text{side 2} + \text{side 3} + \text{side 4}$

$\text{Perimeter}_{(\text{rectangle})} = 2(l) + 2(h) = 2(l + h)$

• The *area of a rectangle* is length times height, or length times width:

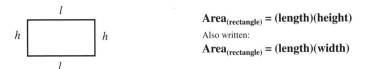

$\text{Area}_{\text{(rectangle)}} = (\textbf{length})(\textbf{height})$

Also written:

$\text{Area}_{\text{(rectangle)}} = (\textbf{length})(\textbf{width})$

• The *area of a rectangle* is also described using the following:

Area rectangle postulate: The area of a rectangle is the length of its base multiplied by the length of its altitude, or height.

$\text{Area}_{\text{(rectangle)}} = (\textbf{base})(\textbf{height})$

• **Example:** You are planning a wedding outside and you want to lay a festive outdoor carpet on the ground and put a small white fence along the edge of the carpet. You've decided it should be in a rectangle of 600 feet by 400 feet. How much fencing and carpet do you need to purchase?

$l = 600 \text{ ft}$

w ⬚ $w = 400 \text{ ft}$

l

If the fencing is going around the edge, or perimeter, of the carpet, then you need to determine the perimeter of the rectangle.

$\text{Perimeter}_{\text{(rectangle)}} = \text{side } 1 + \text{side } 2 + \text{side } 3 + \text{side } 4$, or equivalently,

$\text{Perimeter}_{\text{(rectangle)}} = 2(l) + 2(w) = 2(l + w) = 2(400 \text{ feet} + 600 \text{ feet})$

$= 2(1{,}000 \text{ feet}) = 2{,}000 \text{ feet of fencing}$

You need to purchase the amount of carpet that will cover your 400 foot by 600 foot rectangle, which means you need to determine the area of the rectangle.

$\text{Area}_{\text{(rectangle)}} = (\text{length})(\text{width}) = (600 \text{ feet})(400 \text{ feet}) = 240{,}000 \text{ feet}^2$ or 240,000 square feet

Therefore, you need to purchase 2,000 feet of fencing and 240,000 square feet of carpet!

• **Example:** If the area of a rectangle is 56 square meters and one of its side lengths is 7 meters, what is its perimeter?

One side = 7 meters
Area = 56 square meters

The area of a rectangle is given by: $\text{Area}_{(\text{rectangle})} = (\text{length})(\text{width})$.

If one side length is 7 meters and the area is 56 square meters, this becomes:

56 = (length or width)(7 meters)

(length or width) = 56 meters^2 / 7 meters = 8 meters

The perimeter of a rectangle is given by:

$\text{Perimeter}_{(\text{rectangle})} = 2(l) + 2(w) = 2(l + w)$
Substitute: Perimeter = 2(8 meters) + 2(7 meters)
$= 16 \text{ meters} + 14 \text{ meters} = 30 \text{ meters}$

Therefore, the perimeter of the rectangle is 30 meters.

• **Example:** Suppose you need to put a 5-foot-wide cobblestone walkway around a rectangular office building, which measures 300 by 500 feet. What is the area of cobblestones needed?

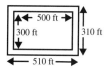

The area of a rectangle is given by: $\text{Area}_{(\text{rectangle})} = (\text{length})(\text{width})$

The area of the building is: $\text{Area}_{(\text{building})} = (500 \text{ ft})(300 \text{ ft}) = 150{,}000 \text{ ft}^2$

Building plus walkway is: $\text{Area}_{(\text{Plus-walk})} = (510 \text{ ft})(310 \text{ ft}) = 158{,}100 \text{ ft}^2$

Area of walkway = (area of building plus walkway) – (area of building)

Area of walkway = (158,100 ft^2) – (150,000 ft^2) = 8,100 ft^2

Therefore, 8,100 square feet of cobblestone is needed.

Area and perimeter of a parallelogram

• A *parallelogram* is a quadrilateral in which both pairs of opposite sides (called **bases**) are parallel to each other.

• The *perimeter of a parallelogram* is the sum of its four sides, or equivalently, two-times the length of one pair of opposite bases (*a*) plus two-times the length of the other pair of opposite bases (*b*).

$$\text{Perimeter}_{\text{(parallelogram)}} = 2a + 2b = 2(a + b)$$

$$\text{Perimeter}_{\text{(parallelogram)}} = 2a + 2b = 2(a + b)$$

• An *altitude of a parallelogram* is a line segment that is perpendicular to both of the opposite bases to which it is drawn. The length of the altitude is called the *height of the parallelogram*. The altitude line may be perpendicular to an extension of a side.

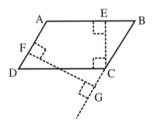

Altitude or height *EC* is perpendicular to bases *AB* and *CD*.

Altitude or height *FG* is perpendicular to bases *AD* and *BC* (extension).

• The *area of a parallelogram* is base times height.

The *area of a parallelogram* is the product of the length of a base (side) and the length of the altitude (height) to that side.

$$\text{Area}_{\text{(parallelogram)}} = \text{(base)(height)}$$

• *Parallelogram area theorem*: The *area of a parallelogram* is the product of the length of a base and the length of its corresponding altitude (height).

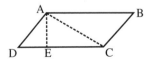

Prove *parallelogram area theorem*:

Given: $\square ABCD$ with base *DC* and altitude *AE*.

Prove: Area $\square ABCD$ = (*DC*)(*AE*).

Proof:

Statements	Reasons
1. $\square ABCD$: base *DC* and altitude *AE*.	1. Given.
2. Draw diagonal *AC*.	2. Two points determine a line.
3. $AB \cong CD$; $AD \cong BC$.	3. Parallelograms have $\cong$ opposite sides.
4. $AC \cong AC$.	4. Identity/reflexive: self = self.
5. $\triangle ACD \cong \triangle CAB$.	5. SSS triangle congruence postulate.
6. Area $\square ABCD$ = area $\triangle ACD$ + area $\triangle CAB$.	6. Area add. post: Area of entire region is sum of areas of non-overlapping regions in it.
7. Area $\square ABCD$ = 2 (area $\triangle ACD$).	7. Substitute #5 and #6.
8. Area $\square ABCD$ = (2)(1/2)(*DC*)(*AE*) = Area $\square ABCD$ = (*DC*)(*AE*).	8. Area of a triangle is (1/2)(base)(height) or Area $\triangle ACD$ = (1/2)(*DC*)(*AE*).

• ***Parallelograms have equal areas*** if they have congruent bases and congruent altitudes, or heights. $(\text{Area}_{(\text{parallelogram})} = (\text{base})(\text{height}))$

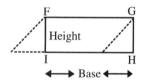

Area $\square ABCD = (CD)(AE)$ Area $\square FGHI = (HI)(FI)$

If $CD \cong HI$ and $AE \cong FI$, then Area $\square ABCD$ = Area $\square FGHI$.

• The area of parallelogram $\square ABCD$ is equivalent to the area of parallelogram $\square ABEF$ because the areas of the two congruent triangles, $\triangle AFD$ and $\triangle BEC$, are equivalent. This results in base *CD* being congruent to base *EF*. In addition, heights *AF* and *BE* are congruent.

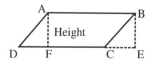

• **Example:** Find the area and perimeter of parallelogram □ABCD if
AD = 4 feet, CD = 5 feet, and the height is 3 feet.

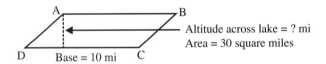

Area$_{(parallelogram)}$ = (base)(height) = (5 ft)(3 ft) = 15 feet^2.

Perimeter$_{(parallelogram)}$ = 2(short side) + 2(long side) = 2(4) + 2(5) = 18 feet.

• **Example:** The approximate area of a man-made, parallelogram-shaped
lake is 30 square miles. The distance across the front side CD is 10 miles.
What is the distance across the lake from side CD to side AB (which
corresponds to the altitude of parallelogram □ABCD).

We can use the equation for the area of a parallelogram and solve for
altitude:

Area$_{(parallelogram)}$ = (base)(altitude)

30 miles^2 = (10 miles)(altitude)

altitude = 30 miles^2/10 miles = 3 miles

Therefore, the distance across the man-made, parallelogram-shaped lake
is 3 miles.

Note that the units miles^2/miles = (miles)(miles)/(miles) = miles because
the one *miles* in the denominator cancelled one of the *miles* in the numerator.

Area and Perimeter of a Rhombus

• A *rhombus* is a *parallelogram* with all four sides of equal length.

• The *perimeter of a rhombus* is the sum of its four sides.

Perimeter$_{(rhombus)}$ = side 1 + side 2 + side 3 + side 4 = (4)(side 1)

- The *area of a rhombus* is the product of a side and the height.

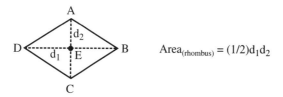

$\text{Area}_{(\text{rhombus})} = (\text{side})(\text{height}) = (AB)(DE)$

This area formula is comparable to the *area of a parallelogram*, which is the product of the length of a base and the length of its corresponding altitude (height).

- Another formula for the area of a rhombus is given by the following:

Area of a rhombus theorem: The *area of a rhombus* equals one-half the product of its diagonals: $\text{Area}_{(\text{rhombus})} = (1/2)d_1 d_2$

$\text{Area}_{(\text{rhombus})} = (1/2)d_1 d_2$

Remember, the *diagonals* of a *rhombus* are perpendicular bisectors of each other and form four congruent triangles.

- Prove the *area of a rhombus theorem*: $\text{Area}_{(\text{rhombus})} = (1/2)d_1 d_2$

$\text{Area}_{(\text{rhombus})} = (1/2)d_1 d_2$

Given: ABCD is a rhombus having diagonals $d_2 = AC$ and $d_1 = DB$.

Prove: Area of rhombus ABCD = $(1/2)d_1 d_2 = (1/2)(AC)(DB)$.

Proof:

$AC \perp DB$ because the diagonals of a rhombus are perpendicular.

AE is the altitude of $\triangle ABD$.

CE is the altitude of $\triangle CBD$.

Area of $\triangle ABD = (1/2)(AE)(DB)$

Area of $\triangle CBD = (1/2)(CE)(DB)$

Area rhombus $ABCD$ = Area of $\triangle ABD$ + Area of $\triangle CBD$

$= (1/2)(AE)(DB) + (1/2)(CE)(DB)$

Factor out $(1/2)(DB)$:

Area rhombus $ABCD = (1/2)(DB)[(AE) + (CE)]$

Substitute $AC = AE + CE$:

Area rhombus $ABCD = (1/2)(AC)(DB) = (1/2)d_1d_2$

• **Example:** If one diagonal of a rhombus is 10 meters and its area is 120 square meters, what is the length of the other diagonal?

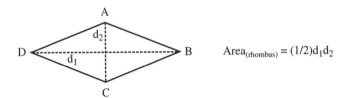

$\text{Area}_{(rhombus)} = (1/2)d_1d_2$

We can solve for the second diagonal by substituting into the area formula: $\text{Area}_{(rhombus)} = (1/2)d_1d_2$

$120 \text{ meters}^2 = (1/2)(10 \text{ meters})(d_1)$

$120 \text{ meters}^2 = (5 \text{ meters})(d_1)$

$d_1 = (120 \text{ meters}^2) / (5 \text{ meters}) = 24 \text{ meters}$

Therefore, the length of the other diagonal is 24 meters.

Area and Perimeter of a Trapezoid

• A *trapezoid* is a quadrilateral having only one pair of opposite sides parallel to each other. These parallel sides are called *bases*. The other pair of opposite sides are not parallel and are called *legs*.

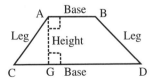

$AB \parallel CD$.
AB and CD are *bases*.
AC and BD are *legs*.
Height $AG \perp AB$ and $AG \perp CD$.

- The ***perimeter of a trapezoid*** is the sum of its four sides.

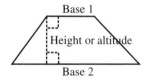

Perimeter$_{(trapezoid)}$ = side 1 + side 2 + side 3 + side 4

- ***Trapezoid area theorem***: The *area of a trapezoid* is the average of its bases multiplied by its height.

Area$_{(trapezoid)}$ = (average of bases)(height)

This is equivalently written:

The area of a trapezoid equals one-half the product of the sum of the lengths of its bases and the length of its altitude (height).

Area$_{(trapezoid)}$ = (1/2)(Base 1 + Base 2)(height)

Area$_{(trapezoid)}$ = (average of bases)(height)
= (1/2)(Base 1 + Base 2)(height)

- Prove the *trapezoid area theorem*:

Area$_{(trapezoid)}$ = (average of bases)(height)

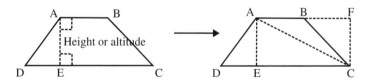

Given: Trapezoid ABCD with bases *AB* and *DC* and altitude *AE*.

Prove: Area trapezoid ABCD = (1/2)(*AE*)(*AB* + *CD*).

Proof:

Statements	Reasons
1. Trapezoid ABCD with bases AB and DC and altitude AE.	1. Given.
2. Draw AC.	2. Two points determine a line.
3. Area ΔACD = (1/2)(DC)(AE).	3. Area of a triangle is (1/2)(base)(height).
4. Draw AF.	4. Extend line AB.
5. Draw CF ⊥ CD.	5. Perpendicular line post: One ⊥ line can be drawn to or through any point on a line.
6. Area ΔACB = (1/2)(AB)(CF).	6. Area of a triangle is (1/2)(base)(height).
7. AE ≅ CF.	7. AB ∥ CD because bases of trap are ∥, and ⊥ segments between ∥ lines are ≅.
8. Area trapezoid ABCD = area ΔACD + area ΔACB = (1/2)(DC)(AE) + (1/2)(AB)(CF).	8. Area add. post: Area of entire region is sum of areas of non-overlapping regions in it.
9. Area trapezoid ABCD = (1/2)(AE)(DC + AB).	9. Substitute #7 and #8 (AE for CF), then factor out (1/2)(AE).

• Remember, the ***median of a trapezoid*** is a line segment that joins the midpoints of the legs and is parallel to the bases. The median of any trapezoid has its length equal to half the sum of the base lengths. Median = (1/2)(Base 1 + Base 2)

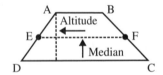

Median $EF = (1/2)(AB + CD)$
$Area_{(trapezoid)}$ = (altitude)(median)
= (1/2)(altitude)(Base 1 + Base 2)
= (1/2)(altitude)(AB + CD)

We can use the ***median*** to calculate the ***area of a trapezoid***.

The area of a trapezoid equals the product of its altitude and its median. Because median = (1/2)(Base 1 + Base 2), then

Area$_{(trapezoid)}$ = (altitude)(median) = (1/2)(altitude)(Base 1 + Base 2)

• **Example:** Find the perimeter and area of trapezoid ABDC.

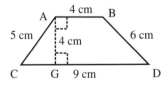

Perimeter$_{(trapezoid)}$ = side 1 + side 2 + side 3 + side 4

= 4 cm + 6 cm + 9 cm + 5 cm = 24 cm

Area$_{(trapezoid)}$ = (average of bases)(height)

= (1/2)(Base 1 + Base 2)(Height)

= (1/2)(4 cm + 9 cm)(4 cm) = 26 cm^2

The perimeter is 24 centimeters and the area is 26 square centimeters.

Area and Perimeter of Other Polygons and Planar Objects

• The *perimeter* is the sum of all side lengths. To determine the *area of polygons and planar objects* that are not triangles, squares, rectangles, parallelograms, or trapezoids, calculate the area of sections of the polygon that form familiar figures, then combine the areas of the sections. Units for area are always squared (square feet, square inches, square miles, square meters, etc.) because of the two dimensions described.

• *Area addition postulate*: The area of a region of a polygon or planar object is the sum of the areas of its non-overlapping parts.

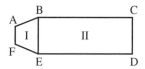

Area ABCDEF = Area I + Area II

= Area trapezoid ABEF + Area rectangle BCDE

• **Example:** Find the area of the shaded region.

The area of the 6 by 6 square is (side)^2, or 6^2 = 36 square units.

There are four quarter-circles centered at points A, B, C, and D, each having 90° within its corner section of the square. The area of a circle is πr^2 (see Section 6.6. *Circumference and Area of Circles and Sectors*), and the square contains four quarter circles, each having a radius of 3.

(4 circles)(1/4 of a circle)(π)(r^2) = (π)(3^2) = 9π square units

The area of the shaded region is the area of the square with four quarter-circles subtracted: (area square) – (area four quarter-circles)

Therefore: Area shaded region = $(36 - 9\pi)$ square units

If we estimate the value of π to be 3.14159, then the area can be written:

Area shaded region $\approx 36 - 9(3.14159) = 7.72569$ square units

Area and Perimeter of Regular Polygons

• A *regular polygon* is a convex polygon that is both equilateral and equiangular. For example, a *square* having four congruent sides and four congruent angles is a *regular quadrilateral*. A *pentagon* having five congruent sides and five congruent angles is a *regular pentagon*.

• *Perimeter of a regular polygon*: The perimeter is the sum of the side lengths. Because the sides of a regular polygon each have the same length, the *perimeter of any regular polygon* can be determined by multiplying the length of one side with the number of sides in the polygon. If the term n-gon is used to represent a polygon with n sides, then perimeter is given by:

Perimeter$_{\text{(regular n-gon)}}$ = n(side)

In other words, if a regular polygon has n sides each of length s, its perimeter is given by:

$$p = ns$$

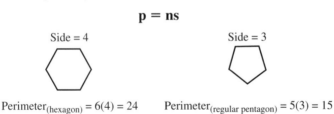

Side = 4   Side = 3

Perimeter$_{\text{(hexagon)}}$ = 6(4) = 24   Perimeter$_{\text{(regular pentagon)}}$ = 5(3) = 15

• The *perimeter of a regular polygon* is also associated with the radius of the polygon and is described using the following theorem:

Perimeter of a regular polygon theorem: The perimeter of a regular polygon having n sides is 2rn sin(180°/n), in which n is the number of sides and r is the length of its radius (which joins the center to any vertex).

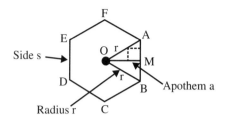

Perimeter$_{\text{(regular n-gon)}}$ = **n(side)** = **2rn sin(180°/n)**

(Remember sin(180°/n) represents the sine of (180°/n). See Section 4.9 *Triangles and Trigonometric Functions.*)

This perimeter formula is true for any number of sides, and can be derived from the relationship between central angle $\angle AOB$, radius r, the two congruent right triangles $\triangle AOM$ and $\triangle BOM$ formed by the apothem, and the side s, (or half side, where $AM = s/2$ and $MB = s/2$).

Because the central angle is $\angle AOB = 360°/n$, and $\angle AOM$ is half its measure, then: $\angle AOM = 180°/n$

The relationship between radius r and side s is found using the trigonometric function for $\angle AOM$ in right triangle $\triangle AOM$:

$\sin(180°/n) = $ opposite/hypotenuse $= AM/r = (s/2)/r = s/2r$, or
$\sin(180°/n) = s/2r$

Solve for s:

$s = 2r \sin(180°/n)$

Substitute for s into the formula for perimeter $p = ns$:

$p = n2r \sin(180°/n)$

• The *area of a regular polygon* can be determined using the apothem and the perimeter.

Area regular polygon theorem: The area of a regular polygon is one-half the product of the length of the apothem and the perimeter.

Area$_{\text{(regular n-gon)}}$ = **(1/2)(apothem)(perimeter)**

Remember the *apothem* of a regular polygon is a line segment that extends from the center of the regular polygon to any side, and is a perpendicular bisector of that side.

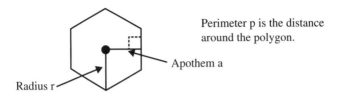

Perimeter p is the distance
around the polygon.

Apothem a

Radius r

Area$_{(regular\ n\text{-gon})}$ = (1/2)(apothem)(perimeter) = (1/2)(a)(p)

Because perimeter = p = ns, where n is the number of sides and s is the length of each side, the area can be expressed as:

Area$_{(regular\ n\text{-gon})}$ = (1/2)(a)(p) = (1/2)(a)(ns)

• The area formula, area$_{(regular\ n\text{-gon})}$ = (1/2)(a)(p) = (1/2)(a)(ns), can be described as calculating the area of the n isosceles triangles in the polygon, where the apothem of the polygon is the altitude of a triangle, and the side of the polygon is the base of each triangle. The area of each isosceles triangle is sectioned off by the radii, and each triangle has the apothem as its altitude and has each side as its base. (Remember, the area of a triangle is: Area = (1/2)(base)(altitude).)

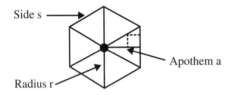

Side s

Apothem a

Radius r

The area of each triangle is:

Area$_{(triangle)}$ = (1/2)(side)(apothem) = (1/2)(s)(a)

The total area of the regular polygon is the sum of the areas of the triangles, or area$_{(triangle)}$ multiplied by number of n sides.

Area$_{(regular\ n\text{-gon})}$ = n × area$_{(triangle)}$ = n(1/2)(s)(a)

• **Example:** In the regular hexagon, if the side length is 20 feet and the apothem is 15 feet, what is the perimeter and area?

20   15

For a regular n-gon: Perimeter$_{\text{(regular n-gon)}}$ = n(side)

For the regular hexagon: Perimeter$_{\text{(hexagon)}}$ = 6(20 feet) = 120 feet

For a regular n-gon: Area$_{\text{(regular n-gon)}}$ = (1/2)(apothem)(perimeter)

For the regular hexagon:

Area$_{\text{(hexagon)}}$ = (1/2)(15 feet)(120 feet) = 900 square feet

Therefore, the perimeter is 120 feet and the area is 900 square feet.

• **Example:** In the regular hexagon with an apothem of $\sqrt{3}$: (a) find the area of the hexagon, and (b) find the areas of the inscribed and circumscribed circles.

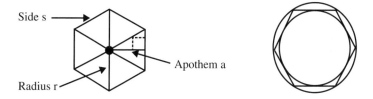

(a)

In a hexagon there are six sides that divide the hexagon into six equal triangles. The six central angles of the hexagon are equal in measure and their values are: c = 360°/n = 360°/6 = 60°.

Because the base angles (along the sides) in each of these triangles are congruent, each base angle must also be 60° (because there are 180° in a triangle). This means that each of the six triangles in a hexagon is an equilateral triangle. Because the side lengths of an equilateral triangle are equal, then a side length of the hexagon equals the radius of the hexagon.

From Chapter 4, we know that an equilateral triangle can be split into two 30:60:90 right triangles. We can use the side ratios of one of the 30:60:90 triangles to find the radius and side length of the hexagon.

The side ratios of a 30:60:90 triangle are leg : leg : hyp = $x : x\sqrt{3} : 2x$, with the half-side as the short leg, the apothem as the long leg, and the radius as the hypotenuse.

Because the long leg (apothem a) is $\sqrt{3}$, then the short leg (half of side s of polygon) is 1, and the hypotenuse (radius r of polygon) is 2. Therefore, the side length of the hexagon is also 2.

Knowing the apothem and side length we can calculate the area:

Area$_{\text{(regular n-gon)}}$ = (1/2)(a)(p) = (1/2)(a)(ns) = (1/2)(6)(2)$\sqrt{3}$ = $6\sqrt{3}$

Therefore, the area of a regular hexagon with an apothem of $\sqrt{3}$ is $6\sqrt{3}$.

(b)

The *radius of the inscribed circle is the apothem*, and the *radius of the circumscribed circle is the radius of the polygon*.

We learn in Chapter 6 that the area of a circle is πr^2, or

$$\text{Area}_{(\text{circle})} = \pi r^2$$

We are given the apothem is $\sqrt{3}$, and because the radius of the inscribed circle is the apothem, then the area of the inscribed circle is:

$$\text{Area}_{(\text{circle})} = \pi r^2 = \pi(\sqrt{3})^2 = 3\pi$$

In part (a), we found that the radius of the polygon is 2, and because the radius of the polygon is also the radius of the circumscribed circle, the area of the circumscribed circle is:

$$\text{Area}_{(\text{circle})} = \pi r^2 = \pi(2)^2 = 4\pi$$

Therefore, the area of the inscribed circle is 3π, and the area of the circumscribed circle is 4π.

5.9 Congruence, Area, and Similarity

Congruence and Area

• *Area congruence postulate*: If two figures are congruent, then they have equal areas.

If Figure A ≅ Figure B, then
area of Figure A = area of Figure B.

• Two figures can have the *same area* and yet *not* be congruent.

Polygons with the same area that are not congruent figures.

• Remember, *similar polygons* have the same shape, but one is larger.

The polygons are similar, having the same shape,
but they are not congruent and do not have the same area.

Similar Polygons

• Cartographers and architects use geometric similarity to produce properly scaled drawings for creating maps and designing structures. For example, one inch on a scale drawing may represent 20 feet of a building or 20 miles on a map.

• If two polygons have the same size and shape, they are called ***congruent polygons***. If two polygons have the same shape, such that their angle measurements are equal and their sides are proportional, however one is larger than the other, they are called ***similar polygons***. The symbol for similar is: ~

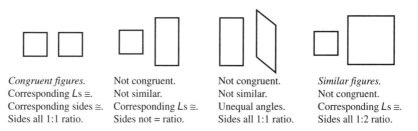

Congruent figures.	Not congruent.	Not congruent.	*Similar figures.*
Corresponding $\angle$s $\cong$.	Not similar.	Not similar.	Not congruent.
Corresponding sides $\cong$.	Corresponding $\angle$s $\cong$.	Unequal angles.	Corresponding $\angle$s $\cong$.
Sides all 1:1 ratio.	Sides not = ratio.	Sides all 1:1 ratio.	Sides all 1:2 ratio.

• For similar polygons:

Corresponding angles are congruent.

Corresponding sides are in proportion, (corresponding side lengths have equal ratios).

• In similar polygons, the ratio of the lengths of corresponding sides is called the *scale factor*.

• In the figure of similar quadrilaterals, we write:

$$A_1B_1C_1D_1 \sim A_2B_2C_2D_2$$

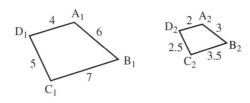

$$\angle A_1 \cong \angle A_2, \quad \angle B_1 \cong \angle B_2, \quad \angle C_1 \cong \angle C_2, \quad \angle D_1 \cong \angle D_2.$$

Corresponding sides are in proportion, and therefore have equal ratios:
$2/4 = 3/6 = 3.5/7 = 2.5/5 = 1/2.$

• In the figure of similar pentagons, we write:

$A_1B_1C_1D_1E_1 \sim A_2B_2C_2D_2E_2$

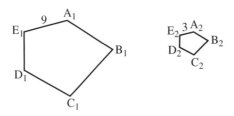

$LA_1 \cong LA_2, LB_1 \cong LB_2, LC_1 \cong LC_2, LD_1 \cong LD_2, LE_1 \cong LE_2.$

$A_2B_2/A_1B_1 = B_2C_2/B_1C_1 = C_2D_2/C_1D_1 = D_2E_2/D_1E_1 = E_2A_2/E_1A_1$

Corresponding sides are in proportion, and therefore have equal ratios: $3/9 = 1/3$.

• *Note*: When you are describing similar polygons, list the corresponding angle vertexes in the same order. For example, if LA_1 corresponds to LA_2, LB_1 corresponds to LB_2, LC_1 corresponds to LC_2, and LD_1 corresponds to LD_2, you should write:

$A_1B_1C_1D_1 \sim A_2B_2C_2D_2$ (not $B_1C_1D_1A_1 \sim A_2B_2C_2D_2$, etc.)

• **Example:** In the figure of similar quadrilaterals, $A_1B_1C_1D_1 \sim A_2B_2C_2D_2$, if $mLA_1 = 110°$, $mLB_1 = 80°$, and $mLC_1 = 100°$, find mLD_2. Also, find side x and the ratio of the perimeters.

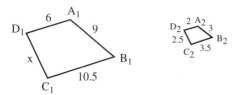

The two similar polygons are quadrilaterals; therefore, the sum of the interior angles is 360°. The sum of the angles in the larger quadrilateral is given by:

$mLA_1 + mLB_1 + mLC_1 + mLD_1 = 360°$

$110° + 80° + 100° + mLD_1 = 360°$

Solve for mLD_1:

$mLD_1 = 360° - 290° = 70°$

Because the quadrilaterals are similar, and corresponding angles are congruent: $mLD_1 = mLD_2 = 70°$

Because the quadrilaterals are similar and corresponding sides are in proportion (side ratios are equal), we can determine side x using the ratio of the sides. The corresponding side ratios are:

2/6 = 3/9 = 3.5/10.5 = 2.5/x = 1/3

Set the x-containing ratio equal to 1/3:

2.5/x = 1/3

x = (2.5)(3) = 7.5

If x = 7.5, then the corresponding side ratio is: 2.5/7.5 = 1/3

The ratio of the perimeters is the same as the ratio of the sides, which is 1/3. Let's prove it:

(2 + 3 + 3.5 + 2.5) / (6 + 9 + 10.5 + 7.5) = 11/33 = 1/3

Therefore, $mLD_2 = 70°$, x = 7.5, and ratio of the perimeters is 1/3.

Similar Regular Polygons

• *Regular polygons* that have the same number of sides are similar.

• Corresponding segments of *similar regular polygons* (having the same number of sides) are in proportion. In similar regular polygons, sides, perimeters, radii, altitudes, and circumferences of inscribed or circumscribed circles are in proportion to each other.

For example, if the ratio of the corresponding *sides* in a regular polygon is 2:1, then the ratio of the corresponding *altitudes* is 2:1, and the ratio of corresponding radii is 2:1.

• *Areas of two similar regular polygons* (having the same number of sides) have ratios as the square of the ratios of corresponding sides (and other corresponding segments between the two polygons).

In other words, the ratio of corresponding areas of two similar regular polygons is the square of the ratio of corresponding sides of the two similar regular polygons. If two similar regular polygons have a side ratio of x:y, then the ratio of their areas is $x^2:y^2$.

5.10 Chapter 5 Summary and Highlights

• A *polygon* is a closed planar figure that is formed by three or more straight line segments that all meet at their endpoints. For a convex polygon with n sides: *Interior angle* sum = (n − 2)180°.

• An *exterior angle* *of a polygon* is an angle formed between the extension of one of its sides and the outside of the polygon. Each exterior angle is a *supplement* of an adjacent interior angle so that: Interior angle + Exterior angle = 180°.

Exterior angles polygon sum theorem: The sum of the measures of the exterior angles, one at each vertex, of any convex polygon equals 360°.

• A *regular polygon* is a convex polygon that is both equilateral and equiangular. Every regular polygon is cyclic. A regular polygon has: *radius* segments that join its center to any vertex, bisecting the *vertex angle*; *apothem* segments extending from its center perpendicular to a side, bisecting the side; *central angles* formed by two radii drawn to two consecutive vertices. For the *angles in a regular polygon* having n sides:

Each interior angle $i = 180°(n − 2)/n$; Each exterior angle $e = 360°/n$;

Interior angle i + exterior angle $e = 180°$; Each central angle $c = 360°/n$.

• *Quadrilaterals* are a four-sided polygons. The interior angles of a quadrilateral sum to 360°.

A *trapezoid* is a quadrilateral with exactly one pair of parallel bases.

A *parallelogram* is a quadrilateral with two pairs of parallel bases and congruent opposite angles.

A *rectangle* is a parallelogram with both pairs of opposite sides parallel and four congruent right angles (equiangular). The minimum definition of a *rectangle* is that it is a parallelogram with one right angle. The diagonals of a rectangle are congruent and bisect each other.

A *rhombus* is a parallelogram having all four sides of equal length and opposite angles congruent (equilateral). The minimum definition of a *rhombus* is that it is a parallelogram in which two adjacent sides are congruent. The diagonals of a *rhombus* are perpendicular bisectors of each other, bisect the vertex angles, and form four congruent *right triangles*.

A *square* is a parallelogram that has all four sides of equal length and four congruent right angles (equiangular and equilateral). The minimum definition of a *square* is that it is a parallelogram with a right angle and two congruent adjacent sides. The diagonals of a square are congruent, bisect each other, are perpendicular, bisect vertex angles, and form four congruent triangles.

• A *parallelogram* is a quadrilateral with the properties: The *opposite sides* are parallel; the *opposite sides* are congruent; the *opposite angles* are congruent; any two *consecutive angles* are supplementary; the *diagonals* bisect each other; and a *diagonal* divides it into two congruent triangles.

• A quadrilateral is a parallelogram if: both pairs of opposite sides are parallel; both pairs of opposite sides are congruent; both pairs of opposite angles are congruent; one pair of opposite sides is parallel and congruent; its diagonals bisect each other; or consecutive interior angles are supplementary.

• *Trapezoid median theorem*: The median of any trapezoid has two properties: It is parallel to both bases, and its length equals half the sum of the base lengths, or median = (1/2)(Base 1 + Base 2).

An *isosceles trapezoid* is a trapezoid in which the non-parallel sides (legs) have equal lengths, the *base angles* are congruent, and the *diagonals* are congruent.

• The *perimeter* of polygons (including squares, rectangles, parallelograms, rhombuses, and trapezoids) is the sum of the lengths of its sides or the distance around. Perimeter = sum of the sides.

• The *area* of a polygon is the measure of its internal size in two dimensions and has units that are squared, or square units.

$$\text{Area}_{(\text{square})} = (\text{side})^2 = (1/2)(\text{diagonal})^2$$

$$\text{Area}_{(\text{rectangle})} = (\text{length})(\text{width}) = (\text{base})(\text{height})$$

$$\text{Area}_{(\text{parallelogram})} = (\text{base})(\text{height})$$

$$\text{Area}_{(\text{rhombus})} = (\text{side})(\text{height}) = (1/2)d_1d_2$$

$$\text{Area}_{(\text{trapezoid})} = (\text{average of bases})(\text{height})$$
$$= (1/2)(\text{Base 1} + \text{Base 2})(\text{height})$$

$$\text{Area}_{(\text{regular n-gon})} = (1/2)(\text{apothem})(\text{perimeter}) = (1/2)(a)(p) = (1/2)(a)(ns)$$

• For similar polygons: Corresponding angles are congruent, and corresponding sides are in proportion (have equal ratios).

• *Regular polygons* that have the same number of sides are similar.

Chapter
6
Circles

Knowing how to make use of the properties of circles can be beneficial when we are studying the world around us. For example, the Greek geographer and astronomer Eratosthenes is thought to be the first person (about 240 B.C.) to successfully measure the ***circumference of the Earth***. The basis for his calculations was the measurement of the *elevation of the Sun* from two different locations. Two simultaneous observations were made, one from Alexandria, Egypt, and the other 5,000 *stadia* away at a site on the Nile near Syene (present Aswan, Egypt). At noon in Syene, when the Sun was directly overhead on the day of the summer solstice, which is the longest day of the year, the Sun's rays beamed down to the bottom of a deep well. At the same time, north of Syene in Alexandria, the Sun's rays shown at a 7.2° (which is 1/50th of a 360° circle) angle from the zenith when measured by the shadow of a pole sticking straight up out of the ground.

Eratosthenes imagined that if the Earth was round, the noonday Sun could not appear in the same position in the sky as seen by two widely separated observers. He therefore compared the angular displacement of the Sun with the distance between the two ground locations. Because the Sun is so far away, it could be assumed that the Sun's rays at the two locations were parallel, and that the difference in the Sun's rays at the two locations was due to the spherical shape of the Earth.

Although the observer at Syene saw the Sun directly overhead at noon, the observer in Alexandria found the Sun was inclined at an angle of 7.2° from vertical. Because a measure of 7.2° corresponds to 1/50th of a full circle (360°), Eratosthenes reasoned that the measured ground distance (between Syene and Alexandria) of 5,000 stadia must represent 1/50th of the Earth's circumference.

Therefore, using the distance of 5,000 stadia between the two locations, and the angle differing by 7.2° (or 1/50th of a 360° circle), the circumference of the Earth could be determined to be 50 times 5,000 stadia, or 250,000 stadia. A stadium is estimated to be equivalent to somewhere between 607 and 738 feet, therefore Eratosthenes' determination was between 29,000 and 35,000 miles. Today's measurements are approximately 24,902 miles at the equator and 24,818 miles at the poles, as data from orbiting Earth satellites show us that the Earth is an oblate spheroid slightly flattened at the poles. (In addition, Earth has an average diameter of approximately 7,918 miles, and an average radius of approximately 3,959 miles.)

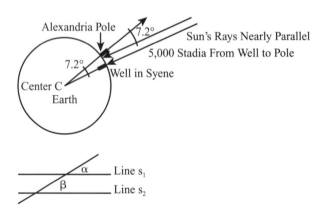

From properties of parallel lines, we know that if lines s_1 and s_2 are parallel, then angles α and β have the same measure.

In Section 6.6, we will learn that if we know the *circumference* C of the Earth, the *diameter of the Earth* d and the *radius* r can be calculated using $C = 2\pi r$ and $d = 2r$. Remember: $\pi = C/d$ *for all circles.*

Another fascinating use of circle properties is that it is possible to estimate the ***diameter of the Sun***. We can do this using the known value of the distance from the Earth to the Sun as 93,000,000 miles, and also that the Sun subtends an angle of 0°31'55" on the surface of the Earth.

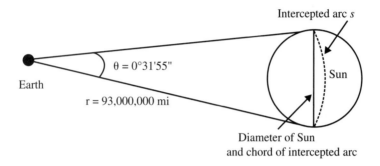

Because the central angle θ is so small relative to the radius r, the arc opposite the central angle (or intercepted arc *s*) and the chord of the intercepted arc (in this case the diameter of the Sun) can be approximated as the same length.

We can use the proportion that the length of the ***arc*** *s* divided by circumference $2\pi r$ is equal to the angle of the arc θ divided by 360°:

$s / 2\pi r = \theta / 360°$

Because $\theta = 0°31'55" = 0.532°$, we can estimate the diameter of the Sun as: $s = 2\pi r\theta / 360° = 2\pi(93{,}000{,}000 \text{ mi})(0.532°) / 360° = 864{,}000$ mi

Central angles, arcs, chords, radius, and diameter are all described in this chapter.

6.1 Circles: Definitions

• A *circle* is a planar shape consisting of a closed curve in which each point on the curve is the same distance from the *center of the circle*. A circle is also described as the set of all points in a plane that are at a given distance (the circle's radius) from a given point in the plane (the circle's center). A circle is often named by its center point, for example circle C in the figure.

• The **radius** of a circle is the distance between the center and any point on the circle. A radius is a line segment that joins the circle's center to a point on the circle. All **radii** drawn for a given circle have the same length. The word radius is used to represent the line segment itself, as well as the length of the radius. The radius is one-half of the diameter.

• A line segment drawn through the center point with its endpoints on the circle is called the **diameter** of the circle. A diameter is a *chord* that includes the center point. The word diameter is used to represent the line segment itself as well as the length of the diameter. The diameter is twice the radius, or (2)(radius) = diameter. A diameter line segment divides a circle into two congruent **semicircles**.

• Any line segment whose endpoints are on the circle is called a **chord** (including the *diameter* line segment). A **chord** is a line segment that joins two points of the circle.

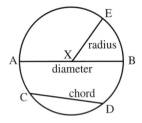

X depicts the point at the center of the circle. *AB* is the diameter chord. *CD* is a chord. *XE*, *XA*, and *XB* are radii. If two chords are congruent, then they are at equal distances from the center of the circle.

• A **secant** is a line that intersects a circle in two points and contains a chord.

• A **tangent line** is *in the plane of a circle* and passes through only one point on a circle, (depicted in preceding figure).

• **One circle per radius postulate**: In a plane, one and only one circle can be drawn for a given radial distance r about any center point.

One circle can be drawn about the
center point C for each radius r_1 and r_2.

• *Congruent circles* are circles that have congruent radii and congruent diameters.

If $r_1 \cong r_2$ and $d_1 \cong d_2$
then circle C $\cong$ circle O.

• If two or more circles lie in the same plane and have the same center point, they are called *concentric circles*. Circles are concentric if and only if they lie in the same plane and have the same center.

Concentric circles

• Points are *concyclic* if and only if they lie on the same circle.

Points A, B, and C lie on the same
circle and are therefore concyclic.

• A circle always measures 360° around, which is equivalent to 2π radians. Half of the circle measures 180°, which is equivalent to π radians. A quarter of a circle measures 90°, which is equivalent to $\pi/2$ radians. The *degrees and radians of a circle* are depicted:

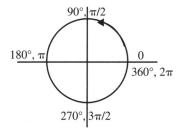

2π *radians* = 360 degrees.

Because there are 360° in a circle, 1° = 1/360th of a circle.

1 *Minute*, denoted by ', is defined as (1/60) of 1°, or approximately 0.0167°.

1 *Second*, denoted by ", is (1/60) of 1 Minute or (1/3600) of a degree or approximately 0.00027778°.

• A *central angle* (vertex at center) subtending an arc equal in length to the radius of a circle is defined as a **radian**. In other words, a *radian* is the measure of the central angle subtending an arc of a circle that is equal to the radius of the circle. Using the definition, 2π radians = 360°:

    1 radian = 360°/2π = 180°/π ≈ 57.296°.

    1 degree = 2π/360° = π/180° ≈ 0.017453 radians.

• **Pi**, or π, defines the ratio between the circumference and the diameter of a circle. More specifically, Pi is equivalent to the circumference divided by the diameter of a circle, Pi = C/d. The value of Pi is approximately 3.141592654 or approximately 22/7.

6.2 Arcs, Central Angles, and Inscribed Angles

• If angles or polygons are drawn inside a circle, any angle whose vertex is at the circle's center point with its sides as radii is called a **central angle**, and any angle whose vertex is on the circle with its sides as chords is called an **inscribed angle**.

Central angle
Vertex is center of circle
and sides are radii of circle.

Inscribed angle
Vertex is on circle and
sides are chords of circle.

• An **arc** is a continuous section of a circle between two points. An arc consists of its endpoints and all points in between. There are three types of arcs: semicircles, minor arcs, and major arcs.

arcAB

• A *semicircle* is an *arc* joining the endpoints of a diameter of a circle. A semicircle arc is labeled with three letters. The first and third are endpoints, and the second lies on the semicircle, such as arcABC (below).

• A *major arc* is larger than a *semicircle* and is named using three letters. The first and third are the endpoints and the second lies on the arc, such as $\overset{\frown}{ACB}$ or arcACB (below).

• A *minor arc* is less than a *semicircle*, and is named using its two endpoints, such as $\overset{\frown}{AB}$ or arcAB (below).

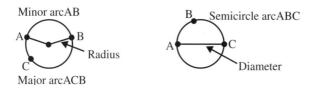

• Sections of circles can be measured by their central angles or by the length of corresponding arcs along the circle's circumference. Arcs can be measured in degrees or in unit length. For example, a 90° arc in a circle with a circumference of 4 inches is 1 inch long, but a 90° arc in a circle with a circumference of 4 yards is 1 yard long.

• The *measure of a minor arc* in degrees is defined as the measure of its corresponding *central angle*. The length of a minor arc is a section of the circumference and is less than half the circumference.

• The *measure of a major arc* in degrees is:

(360° – measure of its minor arc in degrees)

The length of the major arc is a section of the circumference and is greater than half the circumference.

• The measure of a *semicircle* is always 180° or π.

Minor arcAB   Major arcCPD   Semicircle arcEQF

m arcAB = *mL*AOB   *m* arcCPD = 360° – *m* arcCD   *m* arcEQF = 180°
If *mL*AOB = 120°,   If *m* arcCD = 120°, then
then *m* arcAB = 120°.   *m* arcCPD = 360° – 120° = 240°.

• *Arc length*: A section of a circle defined by two or more points is called an *arc*. An arc is measured by its *central angle* or by its *length*, which is *a fraction of the circumference*.

Arc length theorem: The length of an arc whose measure is *m*° is:

arc length = (*m*°/360°)2πr,   where r is the radius of the circle.

In other words, the length of an arc equals the fraction of the circle it covers (*m*°/360°) multiplied by the total distance around the circle (circumference = 2πr).

This can be equivalently expressed as the proportion:

$$\frac{\text{arc length}}{\text{circumference } 2\pi r} = \frac{m°}{360°}$$

• The *measure of a central angle* is proportional to the measure of the arc it intercepts. A central angle is equal in measure to the degree measure of the minor arc that it intercepts. While a central angle is measured by its intercepted arc, a central angle does not equal its intercepted arc, because an angle and an arc are not the same object.

*mL*ABC = *m* arcAC, (but *L*ABC ≠ arcAC)

We can write the proportion for the central angle and arc in the preceding figure as each part is to its whole:

$$\frac{\text{measure } \angle ABC}{360°} = \frac{\text{length of arc}AC}{\text{circumference } 2\pi r} = \frac{\text{area of sector } ABC}{\text{area of circle}}$$

Where the circumference of a circle is $C = 2\pi r$, the area of a circle is area $= \pi r^2$, and $m°$ is the measure of $\angle ABC$.

(Area of circles and sectors is discussed in Section 6.6.)

• **Example:** What is the length of a $36°$ arc in a circle with a 45π circumference?

Because circumference is defined as $2\pi r$ and in this example is equal to the value 45π, then $2\pi r = 45\pi$.

We can solve for arc length by substituting the circumference and $m° = 36°$ into the proportion: (arc length/circumference) $= (m°/360°)$, or arc length $= (m°/360°)$(circumference $2\pi r$)

When $2\pi r = 45\pi$:

arc length $= (36°/360°)(45\pi) = (45/10)\pi = 9\pi/2 \approx 14.14$

Therefore, the length of the arc is $9\pi/2 \approx 14.14$.

• **Example:** What is the radius of a circle if a $30°$ arc has a length of 3π?

 $30°$ arc with a length of 3π.

We can substitute arc length $= 3\pi$ and $m° = 30°$ into the arc length proportion and solve for r:

(arc length/circumference $2\pi r$) $= (m°/360°)$, or
arc length $= (m°/360°)2\pi r$
$3\pi = (30°/360°)2\pi r = (1/12)2\pi r$
$3\pi = (1/6)\pi r$
$18 = r$

Therefore, the length of the radius is 18.

• **Example:** If the radius of a circle is 4 feet, how many degrees are contained in the central angle whose arc has a length of π feet?

 The central angle arc has a length of π feet.
The radius equals 4 feet.

The measure of the central angle equals the measure of the arc it intercepts. We can use the proportion:

(arc length/circumference $2\pi r$) = $(m°/360°)$, where $2\pi r = 2\pi 4$.

Substitute arc length and radius and solve for degrees in central angle:

$(\pi/2\pi 4) = (m°/360°)$
$1/8 = m°/360°$
$m° = 360°/8$
$m° = 45°$

Therefore, the central angle $m°$ is 45°.

• **Example:** In a circle, if an arc measures 4π and the radius is 8 kilometers, what is the corresponding central angle?

 The arc has a length of 4π.
The radius is 8 kilometers.

We can use the proportion: (arc length/circumference $2\pi r$) = $(m°/360°)$, where circumference = $2\pi r = 2\pi 8 = 16\pi$.

Substitute arc length and radius and solve for degrees in central angle:

$m°/360° = 4\pi/16\pi = 1/4$, or $m°/360° = 1/4$
$m° = 360°/4 = 90°$

Therefore, the corresponding central angle is 90°.

• **Example:** In circle C, if arcAB measures 60°, what is the measure of $\angle$B (or $\angle$CBA)?

In the figure, central angle $\angle$C intercepts arcAB. In degrees, we can write: $m\angle C = m$ arcAB = 60°.

Points A, B, and C form a triangle, $\triangle$ABC, with sides $CA \cong CB$ because they are both radii of circle C.

In a triangle, angles opposite congruent sides are congruent: $\angle A \cong \angle B$.

To determine $\angle$B, we can use the equation for the sum of the angles in a triangle, $m\angle A + m\angle B + m\angle C = 180°$, and substitute $m\angle A = m\angle B$ and $m\angle C = 60°$:

$mL\text{B} + mL\text{B} + 60° = 180°$
$2\,mL\text{B} + 60° = 180°$
$2\,mL\text{B} = 120°$
$mL\text{B} = 60°$

Therefore, $\triangle$ABC is an equiangular triangle with 60 degree angles.

• **Example:** If a circle, C, has a radius of 9 meters and is circumscribed about a regular hexagon, what is the length of an arc intercepted by a side of the hexagon?

A regular hexagon has six sides of equal length so the degree measure of each arc of the circle that is intercepted by a hexagon side will equal 1/6 of the total degree measure of a circle, or $(1/6)360° = 60°$.

To determine the arc length, we can use the proportion:

(arc length/circumference $2\pi r$) = $(m°/360°)$,
where $m°$ = degree measure of arc, and
circumference = $2\pi r = 2\pi(9\text{ meters}) = 18\pi$ meters.

Substitute into the proportion and solve for arc length:

(arc length/18π meters) = $(60°/360°)$
arc length = $(1/6)(18\pi$ meters$) = 3\pi$ meters ≈ 9.42 meters

Therefore, the length of an arc intercepted by a side of the hexagon is 3π meters ≈ 9.42 meters.

• *Side note: **Arc length** is also given by:*

Arc length = (radius)(central angle θ *measured in radians*)
= $r\theta$, with θ *measured in radians*
= $(\pi/180°)r\theta°$, with θ the central angle *measured in degrees*

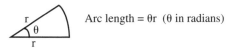

For example, if r = 10 and $\theta = 90° = \pi/2$ radians:

Arc length = $(r)(\theta) = (10)(\pi/2) \approx 15.7$
= $(\pi/180°)r\theta° = (\pi/180°)(10)(90°) \approx 15.7$

Where arc length = $(m°/360°)2\pi r = (\pi/180°)r\theta°$, because $360° = 2\pi$, $180° = \pi$, and $m° = \theta°$.

• *Adjacent arcs* of a circle are arcs that have exactly one endpoint in common.

ArcAB is adjacent to arcBC
and has a common endpoint B.

• *Arc sum postulate*: The measure of the arc formed by two adjacent arcs is the sum of the measures of the two arcs.

In other words, if point B is on an arcABC, then the two arcs formed between A and C, arcAB and arcBC, sum to the total length of arcABC.

If point B is on arcABC, then *m* arcABC = *m* arcAB + *m* arcBC

• **Example:** If *m* arcAB is 75° and *m* arcABC is 125°, what is *m* arcBC?

We can use the *arc sum postulate*, *m* arcABC = *m* arcAB + *m* arcBC, to find *m* arcBC:

$$125° = 75° + m \text{ arcBC}$$
$$m \text{ arcBC} = 125° - 75° = 50°$$

Therefore, *m* arcBC = 50°.

• Point P is a midpoint of arcAPB if arcAP ≅ arcPB. Also, a line, segment, or ray that contains point P bisects arcAPB.

If arcAP ≅ arcPB,
then P is the midpoint of arcAPB
and *OP* bisects arcAPB.

• ***Congruent arcs*** are arcs in the same circle or in congruent circles that have equal measures. Note that if arcs are on two circles that are *not* congruent, they are not congruent arcs and do not have equal lengths, even if the arcs have the same degree measure.

Congruent arcs with equal measure on the same circle (arcAB ≅ arcCD) and on congruent circles (arcAB ≅ arcEF).

ArcAB is not ≅ to arcGH and arcEF is not ≅ to arcGH even though measures are equal.

While an arc may take up 110° of a 360° circle, its *length* will vary with radius length.

• ***Central angle arc theorem***: In the same circle or in congruent circles, two minor arcs are congruent if their corresponding central angles are congruent.

• ***Arc central angle theorem***: In the same circle or in congruent circles, two central angles are congruent if their corresponding minor arcs are congruent.

• **Example:** If $mLAOB$ is 45° and AC and BD are diameters of circle O, what is m arcCD, $mLCOD$, m arcBC, $mLCBO$, and $mLBCO$?

m arcCD:

If $mLAOB = 45°$, then m arcAB $= 45°$ because the measure of a minor arc equals the measure of its corresponding central angle. In addition, if $mLAOB = 45°$, then $mLCOD = 45°$ because $LAOB$ and $LCOD$ are vertical angles and are therefore congruent.

Because $mLCOD = 45°$, then m arcCD $= 45°$.

$mLCOD$:

Because $LAOB$ and $LCOD$ are vertical angles and are congruent, $mLCOD = 45°$.

m arcBC:

Because *m* arcAB = 45° and *AC* is a diameter with its arcABC measuring 180°, then using the arc sum postulate we can determine *m* arcBC:

  m arcAB + *m* arcBC = *m* arcABC
  45° + *m* arcBC = 180°
  m arcBC = 135°

m∠CBO and *m*∠BCO:

∠CBO and ∠BCO are congruent because points BOC form isosceles triangle ∆BOC with sides *OB* and *OC* as radii of circle O and therefore having equal lengths. In a triangle, if two sides are congruent, the angles opposite those sides are congruent so that ∠CBO ≅ ∠BCO.

Because the sum of the angles in any planar triangle is 180°, *m*∠CBO + *m*∠BCO + *m*∠BOC = 180°.

We can determine *m*∠BOC because *m*∠AOB is 45° and *m*∠AOC is 180°; therefore: *m*∠BOC = 180° – 45° = 135°

We can combine the equation for ∆BOC, *m*∠CBO + *m*∠BCO + *m*∠BOC = 180°, with ∠CBO ≅ ∠BCO and *m*∠BOC = 135° to obtain:

  2 *m*∠CBO + 135° = 180°
  2 *m*∠CBO = 45°
  m∠CBO = 22.5°
  m∠CBO = *m*∠BCO = 22.5°

Therefore, *m* arcCD = 45°, *m*∠COD = 45°, *m* arcBC = 135°, and *m*∠CBO = *m*∠BCO = 22.5°.

Inscribed Angles

• An ***inscribed angle*** is any angle whose vertex is on the circle and whose sides are two chords of the circle. An inscribed angle is formed by two chords drawn from the same point on the circle. An inscribed angle intercepts the arc between its sides and is inscribed inside the arc between its sides that contain the vertex.

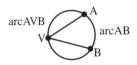

ArcAB is the *intercepted arc* of inscribed ∠AVB.
∠AVB is *inscribed* inside the arcAVB.

∠AVC is said to *intercept* arcAB and to be *inscribed* in arcAVC. For any inscribed angle, the *intercepted arc* and the *inscribed arc* combine to form the circle.

• An **intercepted arc** is the part of a circle that corresponds to an angle and contains the arc between the sides of the angle including its endpoints.

• An inscribed angle is measured by one-half of its intercepted arc.

Inscribed angle theorem: The *measure of an inscribed angle* in a circle is equal in measure to *half of the arc it intercepts*.

$$m\angle ABC = (1/2)(m\ \text{arcAC})$$

• Informal proof of *inscribed angle theorem*:

Given: Circle O with inscribed angle ∠ABC, or ∠B.

Prove: $m\angle B = (1/2)(m\ \text{arcAC})$

To prove we let BC pass through the center of circle O and draw radius OA.

Because all radii are congruent, $OB \cong OA$.

Therefore in △AOB: $\angle A \cong \angle B$.

Because the exterior angle of a triangle equals the sum of the measures of the two remote interior angles: $m\angle A + m\angle B = m\angle AOC$.

Because $\angle A \cong \angle B$, this equation becomes: $m\angle B + m\angle B = m\angle AOC$, or $2\ m\angle B = m\angle AOC$.

Divide both sides by 2: $m\angle B = (1/2)\ m\angle AOC$

Because a minor arc is equal in measure to its central angle:

   $m\ \text{arcAC} = m\angle AOC$

Substitute into the previous equation:

   $m\angle B = (1/2)(m\ \text{arcAC})$

- The following compares arcs of central and inscribed angles:

Central angle
m∠ABC = (*m* arcAC)

Inscribed angle
m∠ABC = (1/2)(*m* arcAC)

- An *inscribed angle* is equal to one-half of the *central angle* formed from the endpoints of the same arc.

In both figures, ∠ABC is an inscribed angle with endpoints on arcAC. Angle ∠ADC is a central angle with endpoints on arcAC. Because they are formed by the same arc, ∠ABC is one-half the measure of ∠ADC or, *m*∠ABC = (1/2) *m*∠ADC, or equivalently, *m*∠ADC = 2 *m*∠ABC.

- **Example:** If ∠B is 35°, what is the measure of arcAC?

Because ∠B, or ∠ABC, is an inscribed angle in arcAC, then:

　　m∠ABC = (1/2)(*m* arcAC)

Substitute ∠B = 35° and solve for *m* arcAC

　　35° = (1/2)(*m* arcAC)

　　70° = *m* arcAC

Therefore, the measure of arcAC is 70°.

• **Example:** If LC is 70° and arcAB is 50°, what is arc x?

Because LC is an inscribed angle in arcABD, then:

   $mLC = (1/2) \, m$ arcABD, or m arcABD $= 2 \, mLC$.

Substitute $LC = 70°$:

   m arcABD $= 2 \, mLC = 2(70°) = 140°$

Because m arcABD $= m$ arcAB $+ m$ arcBD, we can substitute m arcABD $= 140°$ and m arcAB $= 50°$ to find m arcBD $=$ x:

   $140° = 50° + x$

   $90° = x$

Therefore, m arc x $= 90°$.

• *Angle diameter theorem (corollary to inscribed angle theorem)*: If an inscribed angle intercepts a semicircle, it is a right (90°) angle. In other words, if an *inscribed angle* has its sides ending at the endpoints of a *diameter* chord, the vertex of the angle will be a *right angle* (90°), which is one-half of the 180° measurement of the semicircle arc.

 $mLABC = 90° = (1/2)(m$ arcAC$) = (1/2)180°$

• **Example:** If AB is the diameter of the circle, what is the measure of $LACB$, $LADB$, and arcAB?

If an inscribed angle intercepts a semicircle, it is a 90° angle, and angles $LACB$ and $LADB$ both intercept the same semicircle arc, and therefore, each measure 90°. The measure of arcAB is a semicircle, which is 180°.

• *Same arc theorem (corollary to inscribed angle theorem)*: Inscribed angles intercepting the same arc or congruent arcs are equal in measure. In other words, *inscribed angles* with the same endpoints defined by the same arc have the same measure.

If $\angle ADC$ and $\angle ABC$ both intercept arcAC, then $m\angle ADC = m\angle ABC$.

• *Congruent angle theorem (corollary to inscribed angle theorem)*: In the same or congruent circles, congruent inscribed angles have congruent intercepted arcs.

If $\angle C \cong \angle D$, then arcAF $\cong$ arcBE.

If arcAF $\cong$ arcBE, then $\angle C \cong \angle D$.

• *Congruent arc theorem (corollary to inscribed angle theorem)*: In the same or congruent circles, inscribed angles with congruent intercepted arcs are congruent.

• **Example:** If arcAC is 100°, what is the measure of $\angle ABC$ and $\angle ADC$?

By the *inscribe angle theorem*, the measure of an inscribed angle in a circle is equal in measure to *half of the arc it intercepts*.

$m\angle ABC = (1/2)(m \text{ arcAC}) = (1/2)(100°) = 50°$

$m\angle ADC = (1/2)(m \text{ arcAC}) = (1/2)(100°) = 50°$

Therefore, $m\angle ABC = 50°$ and $m\angle ADC = 50°$.

• *Cyclic quadrilateral theorem (corollary to inscribed angle theorem)*: A quadrilateral inscribed in a circle has its opposite angles supplementary. If a quadrilateral is inscribed in a circle, its opposite angles are supplementary.

$mL\text{A} + mL\text{D} = 180°$
$mL\text{B} + mL\text{C} = 180°$

6.3 Chords, Arcs, and Angles

• A *chord* is a line segment that joins two points on a circle.

• If a polygon is inscribed in a circle, the sides of the polygon are chords of the circle.

• *Chord/arc theorem*: In the same circle or in congruent circles, if two chords of a circle are congruent, their associated minor arcs are congruent. In other words, congruent chords have congruent arcs.

If chord $AB \cong$ chord CD,
then arcAB $\cong$ arcCD.

Informal proof:

Given: Chord $AB \cong$ chord CD.

Prove: ArcAB $\cong$ arcCD.

Proof: Draw radii OA, OB, OC, and OD, which are all of equal length.

Therefore, $OA \cong OD$ and $OB \cong OC$.

By SSS congruence postulate, $\triangle$AOB $\cong \triangle$DOC.

Because corresponding parts of congruent triangles are congruent, mLAOB $= mL$DOC.

Because m arcAB $= mL$AOB and m arcCD $= mL$DOC, and also mLAOB $= mL$DOC, then m arcAB $= m$ arcCD, or arcAB $\cong$ arcCD.

• *Arc/chord theorem*: In the same circle or in congruent circles, if two minor arcs of a circle are congruent, their associated chords are congruent. In other words, congruent arcs have congruent chords.

If arc$AB \cong$ arcCD,
then chord $AB \cong$ chord CD.

Proof:

Given: ArcAB $\cong$ arcCD.

Prove: Chord $AB \cong$ chord CD.

Proof: Draw radii OA, OB, OC, and OD, which are all of equal length. Therefore, $OA \cong OD$ and $OB \cong OC$.

Because m arcAB $= m\angle$AOB and m arcCD $= m\angle$DOC, and also arcAB $\cong$ arcCD, then by substitution, $\angle$AOB $\cong \angle$DOC.

By SAS congruence postulate, $\triangle$AOB $\cong \triangle$DOC.

Because corresponding parts of congruent triangles are congruent, chord $AB \cong$ chord CD.

• **Example:** In the figure, determine arc x.

By the *chord/arc theorem*, if chord $AB \cong$ chord CD, then arcAB $\cong$ arcCD. Therefore, x $= 30°$.

• **Example:** In the figure, if $AB \cong CD$ and $AE \cong DF$, prove that $\angle$A $\cong \angle$D.

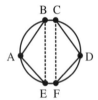

Given: $AB \cong CD$ and $AE \cong DF$.

Prove: $\angle$A $\cong \angle$D.

Proof:

Statements	Reasons
1. Draw BE and CF.	1. Two points determine a line.
2. $AB \cong CD$ and $AE \cong DF$.	2. Given.
3. arcAB $\cong$ arcCD; arcAE $\cong$ arcDF.	3. In a circle, $\cong$ chords have $\cong$ arcs.
4. arcBAE $\cong$ arcCDF.	4. Add. Axiom: Equals added to equals are $=$.
5. $BE \cong CF$.	5. In a circle, $\cong$ arcs have $\cong$ chords.
6. $\triangle$ABE $\cong \triangle$DCF.	6. SSS congruence post: If 3 sides of one $\triangle$ are $\cong$ to 3 sides of other $\triangle$, $\triangle$s are $\cong$.
7. $\angle$A $\cong \angle$D.	7. Corresponding parts of $\cong \triangle$s are $\cong$.

• *Chord/tangent angle theorem*: The measure of an angle formed by a chord and a tangent is equal to one-half the measure of the intercepted arc.

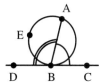

For chord AB and tangent CD:
∠ABC intercepts arcAB, and
$m∠ABC = (1/2)\ m$ arcAB. Also,
∠ABD intercepts arcAEB, and
$m∠ABD = (1/2)\ m$ arcAEB.

• **Example:** If m arcAEB is 200°, find $m∠ABD$.

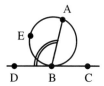

We can use the *chord/tangent angle theorem* to solve for $m∠ABD$:

$m∠ABD = (1/2)\ m$ arcAEB

$m∠ABD = (1/2)(200°) = 100°$

Therefore, $m∠ABD$ is 100°.

• **Example:** If m arcAC is 100° and $m∠A$ is 45°, find $m∠α$.

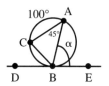

Because $m∠A = (1/2)\ m$ arcBC, or m arcBC $= 2\ m∠A$, then:

m arcBC $= 2\ m∠A = 2(45°) = 90°$

Because m arcAB $+ m$ arcAC $+ m$ arcBC $= 360°$, then:

m arcAB $= 360° - m$ arcAC $- m$ arcBC $= 360° - 100° - 90° = 170°$

Finally, $m∠α = (1/2)\ m$ arcAB $= (1/2)170° = 85°$

Therefore, $m∠α$ is 85°.

• *Perpendicular bisectors of chords* in a circle pass through the center of the circle. A *perpendicular bisector* of any chord drawn toward the center of a circle will pass through the center of the circle. Because of this, we can draw two chords in a circle and draw the perpendicular bisectors of those chords, and where they intersect will be the center of the circle. We can locate the center of the circle using a construction of the intersection of perpendicular bisectors of two chords. (See Chapter 8 Construction 19: Locate the *center* of a *circle*.)

• *Perpendicular bisector chord center theorem*: The perpendicular bisector of a chord passes through the center of the circle.

If *CD* is the perpendicular bisector of chord *AB*, then *CD* passes through the center, O, of the circle.

• *Perpendicular bisector center chord theorem*: If a line passing through the center of a circle is perpendicular to a chord, it also bisects the chord.

If *CD* passes through the center, O, of the circle and is perpendicular to chord *AB*, then it bisects chord *AB*.

How to prove: If we are given that in circle O, *CD* passes through center O and $CD \perp AB$, we can prove that *CD* bisects *AB*.

We can prove this by drawing radii *OA* and *OB* to form congruent right triangles $\triangle OAE$ and $\triangle OBE$. Because corresponding parts of congruent triangles are congruent, $AE \cong EB$, which means *CD* bisects *AB*.

• *Perpendicular bisector chord theorem*: If a line passing through the center of a circle bisects a chord that is not a diameter, it is also perpendicular to the chord.

If *CD* passes through the center O of
the circle and bisects chord *AB*,
then *CD* is perpendicular to *AB*.

• *Diameter chord perpendicular bisector theorem*: A diameter that is
perpendicular to a chord bisects the chord and its arcs.

If diameter *CD* is perpendicular to chord *AB*,
then it bisects chord *AB*, arcACB, and arcADB.

It follows that: $AE \cong EB$, arcAC $\cong$ arcBC,
and arcAD $\cong$ arcBD.

How to prove: If we are given that in circle O, diameter *CD* is perpendicular to chord *AB*, we can prove that $AE \cong EB$ and arcAC $\cong$ arcBC.

We can prove this by drawing radii *OA* and *OB* to form congruent right
triangles $\triangle$OAE and $\triangle$OBE. Because corresponding parts of congruent
triangles are congruent, $AE \cong EB$ and $\angle$AOE $\cong \angle$BOE. Because congruent
central angles have congruent arcs, arcAC $\cong$ arcBC.

• **Example:** If chord *AB* is 3 units from the center of circle O, which has
a radius of 5, what is the length of chord *AB*?

$OE = 3$ and radius $= 5$

In the figure, *OE* falls on diameter chord *CD*, which is a perpendicular
bisector to chord *AB*.

We are given that $OE = 3$ and $OA = 5$, which form two sides of right
triangle $\triangle$AOE.

To determine length *AB*, we can first find *AE*, the unknown long leg of
right triangle $\triangle$AOE, using the Pythagorean Theorem or recognizing that
it must be a 3:4:5 right triangle. Side *AE* is therefore equal to 4. Because
OE bisects *AB*, $AE \cong EB = 4$, so that $AE + EB = AB = 8$.

Therefore, the length of chord *AB* is 8.

• ***Congruent distance/chords theorem***: In the same circle or in congruent circles, if chords are of equal distance from the center of the circle, they are congruent.

• ***Congruent chords/distance theorem***: In the same circle or in congruent circles, if chords are congruent, they are of equal distance from the center of the circle.

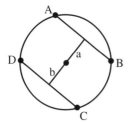

If length a = length b,
then chord $AB \cong$ chord CD.
If chord $AB \cong$ chord CD,
then length a = length b.

How to prove: If we are given that in circle O, chord $AB \cong$ chord CD, $OE \perp AB$, and $OF \perp CD$, we can prove that $OE \cong OF$.

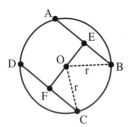

We can prove this by drawing radii OC and OB to form congruent right triangles $\triangle OBE$ and $\triangle OCF$. Because corresponding parts of congruent triangles are congruent, $OE \cong OF$.

• **Example:** If chord $AB = 8$, length OE is 3, and $AB \cong CD$, find the radius r of the circle.

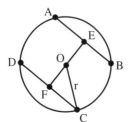

$AB = 8$, $OE = 3$, and $AB \cong CD$.

If chord $AB \cong$ chord CD, then length $OE =$ length $OF = 3$.

We also know that the line that OE and OF lie on passes through the center of the circle along a diameter and that by the *diameter chord perpendicular bisector theorem*, a diameter that is perpendicular to a chord bisects the chord and its arcs. From this, we know that EF is perpendicular to both CD and AB, and that $AE \cong EB$ and $CF \cong FD$. Radius r forms a right triangle with segments OF and CF. This right triangle has leg $OF = 3$, leg $CF = 8/2 = 4$, and hypotenuse r. We can use the Pythagorean Theorem to determine r or recognize that this right triangle is a 3:4:5 Pythagorean triplet.

Therefore, r = 5.

• ***Intersecting chords theorem***: If two chords intersect inside a circle, the product of the lengths of the segments of one chord is equal to the product of the lengths of the segments of the other chord.

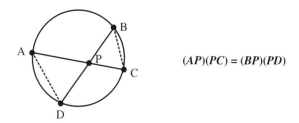

$(AP)(PC) = (BP)(PD)$

How to prove: If we are given that AC and BD intersect at P, we can prove that $(AP)(PC) = (BP)(PD)$.

We can prove this by drawing chords AD and BC. Because $\angle A \cong \angle B$ and $\angle C \cong \angle D$ (inscribed angles that intercept the same arc are congruent), $\triangle APD \sim \triangle BPC$. The ratios of the corresponding sides of similar triangles are equal (corresponding sides are in proportion) so that $(AP)/(BP) = (PD)/(PC)$, which rearranges by cross-multiplication to: $(AP)(PC) = (BP)(PD)$.

• **Example:** In the above figure, if $AP = 8$, $BP = 6$, and $PD = 7$, what is PC?

Using the intersecting chords theorem $(AP)(PC) = (BP)(PD)$, we can substitute and solve for PC:

$(8)(PC) = (6)(7) = 42$, or $(8)PC = 42$

$PC = 42/8 = 5.25$

Therefore, PC is 5.25.

6.4 Secants, Angles, Arcs, and Segments

• A *secant* is a line that intersects a circle at two points. A secant line contains a chord (which is a segment whose endpoints lie on a circle).

• A *secant angle* is an angle whose sides are contained within two secants of a circle. Each side of a secant angle intersects the circle in at least one point other than the angle's vertex.

• *Secant (chord) angle in circle theorem*: A *secant angle* whose vertex is inside a circle is equal in measure to half the sum of the arcs intercepted by it and its vertical angle. This theorem is also true for *chords*.

$mLAVD = (1/2)(m \text{ arcAD} + m \text{ arcBC})$
$mLAVB = (1/2)(m \text{ arcAB} + m \text{ arcCD})$

Proof:

Given: Secants *AC* and *BD* intersect in secant angle *LAVB* in a circle.

Prove: $mLAVB = (1/2)(m \text{ arcAB} + m \text{ arcCD})$.

Proof:

Statements	Reasons
1. Secants *AC&BD* forming *LAVB*.	1. Given.
2. Draw *BC* forming ΔVBC.	2. Two points determine a line.
3. *mLAVB* = *mLVBC* + *mLVCB*.	3. Measure of exterior *L* of a Δ equals sum of measures of Δ's two remote interior *L*s.
4. *mLVBC* = (1/2) *m* arcCD; *mLVCB* = (1/2) *m* arcAB.	4. Measure of an inscribed *L* equals one-half the measure of its intercepted arc.

5. *mLAVB* = (1/2) *m* arcCD
   + (1/2) *m* arcAB
   = (1/2)(*m* arcCD + *m* arcAB).
	5. Substitution of #3 and #4.

• **Example:** If $mLAVD$ is 30° and m arcAD is 24°, find $mLBVC$ and m arcBC.

Because $LAVD$ and $LBVC$ are vertical angles, they have equal measure: $mLAVD = mLBVC = 30°$.

To find m arcBC we can use the *secant (chord) angle in circle theorem*, $mLAVD = (1/2)(m$ arcAD $+ m$ arcBC), and solve for m arcBC.

$30° = (1/2)(24° + m$ arcBC$) = (1/2)(24°) + (1/2)(m$ arcBC$)$

$30° = 12° + (1/2)(m$ arcBC$)$

$18° = (1/2)(m$ arcBC$)$

$36° = m$ arcBC

Therefore, $mLBVC = 30°$ and m arcBC $= 36°$.

• *Secant angle outside circle theorem*: A *secant angle* whose vertex is outside a circle is equal in measure to one-half the (positive) difference between the measures of the arcs intercepted by it. This is true for two tangents or a secant and a tangent.

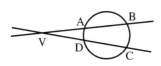

Two secants   Two tangents   Secant and tangent

$mLBVC = (1/2)(m\text{ }arcBC - m\text{ }arcAD)$

How to prove: If we are given secants AB and CD intersecting in secant angle $LBVC$ outside the circle, we can prove that $mLBVC = (1/2)(m$ arcBC $- m$ arcAD$)$.

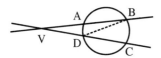

We can prove this by drawing chord BD to create $\triangle VBD$. Because the measure of the exterior angle of a triangle equals the sum of measures of the two remote interior angles: $mLBVC + mLVBD = mLBDC$

Rearranging: $mLBVC = mLBDC - mLVBD$

Because the measure of an inscribed angle equals one-half the measure of its intercepted arc:

$m\angle$BDC = (1/2) m arcBC and $m\angle$VBD = (1/2) m arcAD

By substitution: $m\angle$BVC = (1/2) m arcBC – (1/2) m arcAD

= $m\angle$BVC = (1/2)(m arcBC – m arcAD)

• **Example:** If m arcAD is 20° and m arcBC is 40°, determine $m\angle$BVC.

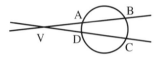

To find $m\angle$BVC we can use the *secant angle outside circle theorem*:

$m\angle$BVC = (1/2)(m arcBC – m arcAD)
$m\angle$BVC = (1/2)(40° – 20°) = 10°

Therefore, $m\angle$BVC = 10°.

• **Example:** If m arcAB is 125°, m arcBC is 50°, and m arcCD is 160°, find $m\angle$BVC.

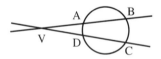

To find $m\angle$BVC, we first determine m arcAD by subtracting known arcs from the 360 degree circle:

360° = 125° + 50° + 160° + m arcAD
m arcAD = 360° – (125° + 50° + 160°) = 360° – 335°
m arcAD = 25°

Next, we can use the *secant angle outside circle theorem*:

$m\angle$BVC = (1/2)(m arcBC – m arcAD)
$m\angle$BVC = (1/2)(50° – 25°) = 12.5°

Therefore, $m\angle$BVC = 12.5°.

• If a line segment intersects a circle in two points, and one of those points is the endpoint of the segment, then the segment is a *secant segment* to the circle.

• *Secant segments theorem*: If two secant segments are drawn to a circle from an external point and intersect outside the circle, the *product of the lengths* of one (entire) secant segment and its external segment is equal to the product of the lengths of the other (entire) secant segment and its external segment. In the figure, $(VA)(VB) = (VD)(VC)$.

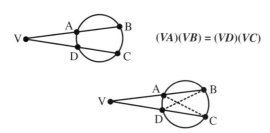

$(VA)(VB) = (VD)(VC)$

How to prove:

If we are given that secant segments VB and VC intersect at external point V, we can prove that $(VA)(VB) = (VD)(VC)$.

We can prove this by first drawing chords AC and BD.

Because $\angle V \cong \angle V$ and also $\angle B \cong \angle C$ (inscribed angles, $\angle ABD$ and $\angle ACD$, that intercept the same arcAD are congruent), then $\triangle VAC \sim \triangle VDB$ by AA Similarity. The ratios of the corresponding sides of similar triangles are equal so that $(VA)/(VD) = (VC)/(VB)$, which rearranges by cross-multiplication to: $(VA)(VB) = (VD)(VC)$.

• **Example:** In the following figure find x.

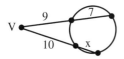

We can use the secant segments theorem and solve for x:

$$(9)(9 + 7) = (10)(10 + x)$$
$$144 = 100 + 10x$$
$$44 = 10x$$
$$4.4 = x$$

Therefore, segment x = 4.4.

• *Secant tangent segments theorem*: If a secant segment and a tangent segment are drawn to a circle from an external point, the *product of the lengths* of the (entire) secant segment and its external segment is equal to the square of the tangent segment.

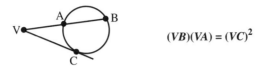

$$(VB)(VA) = (VC)^2$$

How to prove: If we are given that VB and VC intersect at V, we can prove that $(VB)(VA) = (VC)^2$.

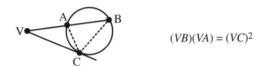

$$(VB)(VA) = (VC)^2$$

We can prove this by first drawing chords AC and BC.

Because $\angle V \cong \angle V$ and also $\angle B \cong \angle VCA$ (angles intercept the same arcAC), $\triangle VBC \sim \triangle VCA$.

The ratios of the corresponding sides of similar triangles are equal so that $(VC)/(VA) = (VB)/(VC)$, which rearranges by cross-multiplication to: $(VB)(VA) = (VC)^2$.

- **Example:** In the following figure determine x.

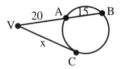

We can calculate x using the *secant tangent segments theorem*:

$$(VB)(VA) = (VC)^2$$
$$(20 + 15)(20) = (x)^2$$
$$700 = (x)^2$$
$$x = \sqrt{700} \approx 26.46$$

Therefore, $x \approx 26.46$.

6.5 Tangents

- A *tangent line* is a line *in the plane of a circle* that passes through only one point on a circle. The point where the tangent line intersects the circle is called the *point of tangency*. If a radius line segment is drawn from the center of the circle to a point of tangency, the tangent line and the radius line segment will be perpendicular to each other.

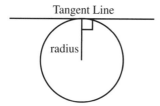

Tangent Line

radius

• ***Tangent to radius theorem***: If a line is tangent to a circle, then the line is *perpendicular* to the radius drawn to the point of tangency.

• ***Converse of tangent to radius theorem***: If a line in a plane of a circle is *perpendicular* to the outer endpoint of the radius of the circle, then it is tangent to the circle.

Also, see Construction 20 in Chapter 8.

• ***Diameter to tangent theorem***: In a circle, if a tangent line meets a diameter at the point of tangency, then they are *perpendicular* to each other.

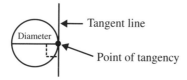

Diameter

Tangent line

Point of tangency

• ***Tangent to diameter theorem***: If a chord in a plane of a circle is perpendicular to a tangent at the point of tangency, the chord is a diameter.

• **Example:** How would you find the center of a circle using the *tangent to diameter theorem*?

We first draw two tangents to the circle. Then, at the point of tangency of each tangent line, draw a chord perpendicular to each tangent. Because of the *tangent to diameter theorem*, these two chords will be diameters and will intersect each other at the center of the circle.

• If a line is tangent to a circle, then any segment of the tangent line having the point of tangency as its endpoint is a tangent segment to the circle.

  AB and *BC* are both tangent segments to the circle.

• ***Tangent segments theorem***: The *tangent segments* to a circle from an external point are congruent. In other words, if two tangent segments intersect outside a circle, then the tangent segments have equal lengths.

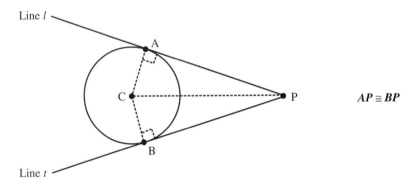

AP ≅ *BP*

Also see Construction 21 in Chapter 8.

How to prove tangent segments theorem: If we are given in circle C tangent segments *AP* and *BP*, we can prove that *AP* ≅ *BP*. We can prove this by drawing radii *CA* and *CB* and segment *CP*. Because tangents are perpendicular to the radius at the point of tangency, *AP* ⊥ *CA* and *BP* ⊥ *CB* so ∠A ≅ ∠B and they are right angles. In addition, *CA* ≅ *CB* (both radii) and *CP* ≅ *CP* so by HL, hypotenuse-leg, ΔPAC ≅ ΔPBC. Because corresponding parts of congruent triangles are congruent, *AP* ≅ *BP*.

• **Example:** In the above figure of the two tangents *AP* and *BP* drawn from an external point P, if *AP* = 5,000,000 miles, what is *BP*?

Because *AP* ≅ *BP*, *BP* is also 5,000,000 miles.

• A *common tangent* is a line that is tangent to two coplanar circles.

• A *common internal tangent* is a line that is tangent to two coplanar circles and that *intersects* the segment joining the centers of the two circles.

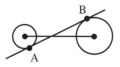

Line *AB* is a *common internal tangent.*

• A ***common external tangent*** is a line that is tangent to two coplanar circles and does *not intersect* the segment joining the centers of the two circles.

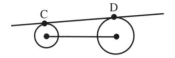

Line *CD* is a *common external tangent.*

• ***Tangent circles*** are two coplanar circles that are tangent to the same line at the same point.

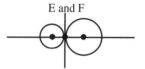

Circles E and F are   Circles G and H are
externally tangent.   *internally tangent.*

• **Example:** In the figure the two circles, circle O and circle Q, are tangent externally and have tangents *PA*, *PB*, and *PC* drawn to them, with *PB* and *PA* tangent to circle O and *PC* and *PA* tangent to circle Q. Prove that tangents *PB* and *PC* drawn to the circles O and Q from a point P on their common internal tangent are of equal length.

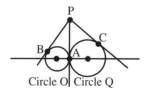

Circle O   Circle Q

Given: Circle O and circle Q with *PA* as the common internal tangent; *PB* and *PC* are external tangents to circles O and Q from P.

Prove: $PB \cong PC$.

Proof:

Statements	Reasons
1. *PB* and *PA* are tangents to circle O.	1. Given.
2. *PB* ≅ *PA*.	2. Tangent segments drawn to a circle from an external point are ≅.
3. *PC* & *PA* are tangents to circle Q.	3. Given.
4. *PC* ≅ *PA*.	4. Tangent segments drawn to a circle from an external point are ≅.
5. *PB* ≅ *PC*.	5. Transitive: Substitute #2 and #4.

6.6 Circumference and Area of Circles and Sectors

• *Sections of circles* can be measured by their central angles or by the length of corresponding arcs along the circle's circumference. Arcs can be measured in degrees or in unit length. A complete circle is 360° or 2π radians.

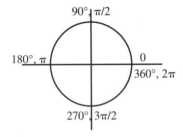

360 degrees = 2π *radians*

• The *perimeter of a circle* is called the **circumference**.

Perimeter of a circle = circumference = C = $2\pi r$ = πd where r = radius, d = diameter, and $\pi \approx 3.141592654...$

r = 2 in.

If r = 2, the circumference $2\pi r$ = 4π or approximately 12.57.

• In any circle, the circumference divided by the diameter is always Pi, or π, which is an irrational number with endless non-repeating digits to the right of the decimal point. π ≈ 3.141592654...

• The **circumference** *of a circle* can also be defined as the limit of the perimeters of the *inscribed regular polygons* it can contain. As the number of sides in the inscribed polygon increases, the polygon begins to be shaped more like the circle, and the perimeter of the many-sided polygon becomes an increasingly close approximation to the circumference of the circle. As the number of sides increase, and the length of each side gets smaller, the *apothem* and the *radius* of the polygon approach the same length, which is the radius of the circle.

Square   Hexagon   Octagon

We can represent this mathematically using one of the perimeter formulas for a regular polygon, and increasing the number of sides until the value approaches the equation: Circumference = 2πr.

Remember the perimeter formula in Section 5.8:

Perimeter$_{\text{(regular n-gon)}}$ = n(side) = 2rn sin(180°/n)

Substitute number of sides n until 2rn sin(180°/n) approaches 2rπ:

Perimeter$_{\text{(regular n-gon)}}$ = 6(side) = 2r6 sin(180°/6) = 2r(3.0000000...)

Perimeter$_{\text{(regular n-gon)}}$ = 24(side) = 2r24 sin(180°/24) = 2r(3.132627...)

Perimeter$_{\text{(regular n-gon)}}$ = 100(side) = 2r100 sin(180°/100) = 2r(3.1410759...)

Perimeter$_{\text{(regular n-gon)}}$ = 1,000(side) = 2r1,000 sin(180°/1,000) = 2r(3.1415875...)

Perimeter$_{\text{(regular n-gon)}}$ = 10,000(side) = 2r10,000 sin(180°/10,000) = 2r(3.1415926...)

We can see that this is converging to the value of the *circumference of a circle* = 2rπ = 2r(3.141592654...)

• **Circumference theorem**: The circumference C of a circle with radius r is:

C = 2πr

Because diameter d equals two-times the radius r, circumference C:

C = 2πr = πd

- *Side note*: Because C = πd, **Pi** = π = C/d = circumference/diameter.

- **Example:** If the circumference of an extra large circular pizza is 6 feet, determine its diameter and area.

Because circumference C = 2πr = πd, we can solve for d:

C = πd

d = C/π = 6 feet/π ≈ 1.91 feet

Therefore, the diameter of the pizza is 6 feet/π ≈ 1.91 feet

In the following paragraphs we learn that the area of a circle = $\pi r^2 = \pi(d/2)^2$.

We can substitute the diameter d = 1.91:

Area of pizza = $\pi(d/2)^2 = \pi(d^2/2^2) = \pi(d^2/4) = (d^2)(\pi/4)$
= $(1.91\ \text{feet})^2(\pi/4) = (3.65\ \text{feet}^2)(\pi/4) \approx 2.87\ \text{feet}^2$

Therefore, the area of the pizza is approximately 2.87 square feet.

Area of a Circle

- The **area of a circle** with radius r, diameter d, and π ≈ 3.14 is:

    Area of a circle = $\pi r^2 = \pi(d/2)^2$

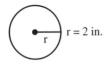

r = 2 in.

In this figure, area = $\pi r^2 = \pi(2\ \text{inches})^2 = \pi4\ \text{inches}^2 \approx 12.56$ square inches

- Using a similar process as we did for *circumference*, we can also define the **area of a circle** as the limit of the areas of the *inscribed regular polygons* it can contain. As the number of sides in the inscribed polygon increases, the polygon begins to be shaped more like the circle, and the *area* of the many-sided polygon becomes an increasingly close approximation to the area of the circle.

Radius
Apothem

Square

Hexagon

Octagon

The area formula for a regular polygon is:

$$Area_{(regular\ n\text{-}gon)} = (1/2)(a)(p)$$

where a is the *apothem* and p is the *perimeter.*

As the number of sides of the inscribed polygon increases, and the length of each side gets smaller, the apothem and the radius approach the same length so that a = r. In a circle, the perimeter (circumference) is $2\pi r$. If we substitute radius for apothem and $2\pi r$ for perimeter, we obtain:

$$Area_{(regular\ n\text{-}gon\ \rightarrow circle)} = (1/2)(a)(p) = (1/2)(r)(2\pi r) = \pi r^2.$$

• *Circle area theorem*: The area of a circle, in which r is the length of the radius, is: Area $= \pi r^2$.

Remember, *Pi*, or π, defines the ratio between the circumference and the diameter of a circle. More specifically, Pi = circumference ÷ diameter. The value of Pi is approximately 3.141592654...

• **Example:** Determine the circumference of a circle if the area is 36π square meters.

First, find r using the area formula and then use r to calculate circumference.

$$Area_{(circle)} = \pi r^2$$
$$36\pi\ meters^2 = \pi r^2$$
$$36\ meters^2 = r^2$$

Take the square root of each side:

r = 6 meters

Circumference $= 2\pi r = 2\pi(6\ meters) = 12\pi\ meters$

Therefore, the circumference is 12π, or approximately, 37.699 meters.

• **Example:** You have a circular piece of wood and you need to create four identical circles, each as large as possible, from the original piece. What is the radius of each of the new circles given the area of the original larger circle is 36π square meters. Also, compare the area of the four small circles with the area of the large 36π square meter circle.

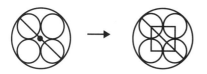

Let radius of small circles be x, then side of square is 2x and diagonal of square is $2x\sqrt{2}$.

We can represent the radius of each of the four small congruent circles as x. If we connect the centers of each of the four small circles, a square is formed with side lengths of 2x and a diagonal of $2x\sqrt{2}$ (because the measure of the hypotenuse of an isosceles right triangle is $\sqrt{2}$ multiplied by the measure of a leg).

The diameter of the large circle is the diagonal of the square plus a radius from two of the smaller circles:

$2x + 2x\sqrt{2}$ = diameter of large circle

The radius of the larger circle is half the diameter, or

$r_{large\ circle} = x + x\sqrt{2} = x(1 + \sqrt{2})$

If the area of the large circle is 36π square meters, and the area of a circle is πr^2, then:

$\pi r^2 = 36\pi$, or $r^2 = 36$

Take the square root of both sides:

$r_{large\ circle}$ = 6 meters, which can be written: $r_{large\ circle} = x(1 + \sqrt{2}) = 6$

Now we can solve for the radius x of the small circles using $r_{large\ circle}$:

$r_{large\ circle} = x_{small\ circle}(1 + \sqrt{2}) = 6$

$x_{small\ circle} = 6/(1 + \sqrt{2})$, which is approximately 2.49 meters.

Therefore, the radius of each small circle is approximately 2.49 meters.

Finally, the area of the four small circles is
$(4)\pi r^2 = (4)\pi(2.49)^2 = 24.8\pi$ square meters.

The area of the large circle is $36\pi \approx 113.1$ square meters and the area of the four small circles is $24.8\pi \approx 77.9$ square meters, which means we wasted approximately 35.2 square meters of wood.

Area of a Sector

• A *sector of a circle* is a region of the circle extending from the center point between two radii to an arc of the circle between the two endpoints of the radii. A sector is bounded by two radii and the arc they encompass. In the figure, sector AOB is a region of circle O bounded by arcAB and the two radii *OA* and *OB* with their endpoints as the endpoints of arcAB. ArcACB is also a sector of circle O.

Sectors in a pie chart are often used in finance and marketing for strategy and to analyze data.

• The ***area of a sector*** of a circle is a *fraction of the area of the whole circle*. Similarly, the ***arc of a given sector*** is a *fraction of the circumference of a circle*.

• ***Sector area theorem***: The area of a sector whose arc has a measure of $m°$ is given by $(m°/360°)\pi r^2$, in which r is the radius of the circle. In other words, the area of a sector whose arc has a measure of $m°$ equals the fraction of the circle it consumes $(m°/360°)$, multiplied by the total area of the circle (πr^2).

For a circle with its central angle = $\angle$ABC, the following proportion is true:

$$\frac{\text{measure } \angle ABC}{360°} = \frac{\text{length of arcAC}}{\text{circumference}} = \frac{\text{area of sector ABC}}{\text{area of circle}}$$

The circumference of a circle is: $C = 2\pi r$.

The area of a circle is: Area $= \pi r^2$.

If central angle $\angle$ABC $= 60°$, then it is $60°/360°$, or 1/6th of the circle.

We can find the *length of the arc* of a sector using the measure of the central angle. If the central angle contained in a sector is 1/6 of the measure of the entire circle (which is always $360°$), then the measure of the arc (arcAC) of that sector will be 1/6 of the circumference of the circle. Therefore, the length of arcAC is 1/6th of the circumference of the circle.

The area of sector ABC is 1/6th of the area of the circle, or:

   (1/6) area of circle = (1/6) πr^2.

• To determine the area of a sector of a circle given the sector's central angle and the circle's radius, we can use the proportion:

$$\frac{\text{measure of central angle}}{360°} = \frac{\text{area of sector}}{\text{area of circle } \pi r^2}$$

For a central angle (LABC) of 60° and a radius of 12 feet with the area of a circle given by πr^2, then the proportion becomes:

60°/360° = (area of sector)/πr^2

60°/360° = (area of sector)/$\pi (12)^2$

1/6 = (area of sector)/144π

(1/6)144π = area of sector

area of sector = 144π/6 = 24π square feet

Therefore, the area of the sector is 24π square feet.

• **Example:** Determine the area of a sector given its associated central angle is 90°, and the length of its arc is 12π centimeters.

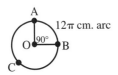

We can first calculate the radius of the circle using the proportion:

$$\frac{\text{measure of central angle}}{360°} = \frac{\text{arc length}}{\text{circumference of circle } 2\pi r}$$

90°/360° = (12 cm.)π/2πr

Reduce each side and solve for r:

1/4 = 6/r

r = 24 cm.

Knowing the radius, r = 24 cm., we can calculate the area of the circle as:
Area of circle = πr^2 = $\pi(24 \text{ cm.})^2$ = 576π cm.^{2}

We can now use the following proportion to calculate the area of the sector:

$$\frac{\text{measure of central angle}}{360°} = \frac{\text{area of sector}}{\text{area of circle}}$$

90°/360° = (area of sector)/576π cm.^{2}

Reduce and solve for sector area:

1/4 = (area of sector)/576π cm.^{2}

area of sector = 576π cm.^{2}/4

area of sector = 144π cm.^{2}

Therefore, the area of the sector is 144π square centimeters.

• **Example:** Calculate the area of sector ABC if ∠B = 90° and r = 10 in.

We can use the following proportion to calculate the area of the sector:

$$\frac{\text{measure of central angle}}{360°} = \frac{\text{area of sector}}{\text{area of circle}}$$

$(90°/360°) = (\text{area of sector})/(\pi)(10 \text{ in.})^2$
$(1/4) = (\text{area of sector})/(\pi 100 \text{ in.}^2)$
$\text{area of sector} = (1/4)(\pi 100 \text{ in.}^2) = \pi 25 \text{ in.}^2 \approx 78.54 \text{ in.}^2$

Therefore, the area of sector ABC is $\pi 25 \text{ in.}^2 \approx 78.54$ square inches.

Segment of a Circle

• A *segment of a circle* is a region of the circle bounded by a chord and its arc.

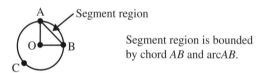

Segment region is bounded by chord *AB* and arc*AB*.

• The ***area of a minor segment of a circle*** equals the area of its sector minus the area of the triangle formed by its chord and radii.

• **Example:** In the above figure, if the radius is 6 and the central angle containing sector AOB is 90°, determine the area of the segment.

We can use the proportion:

$$\frac{\text{measure of central angle}}{360°} = \frac{\text{area of sector}}{\text{area of circle } \pi r^2}$$

Area of the sector $= (90°/360°)\pi r^2 = (90°/360°)\pi 6^2 = (1/4)36\pi = 9\pi$
The area of right $\triangle$AOB is $(1/2)(\text{base})(\text{height}) = (1/2)(6)(6) = 18$
Area of segment = area of sector – area of triangle = $9\pi - 18 \approx 10.26$.

6.7 Circumscribed and Inscribed Polygons

• In Chapter 5, Section 5.3, we learned about *regular polygons*, which are convex polygons that are both equilateral and equiangular.

Each figure has congruent angles and congruent sides.

• An *equilateral polygon* inscribed inside a circle is a *regular polygon*.

• **Regular cyclic polygon theorem**: Every *regular polygon* is cyclic.

If a circle is drawn through the vertices of a regular polygon, the distance from the center to each vertex will be the same.

A circle can be *circumscribed* around any regular polygon.

A circle can be *inscribed* inside any regular polygon.

The *center of a circle* circumscribed about a regular polygon is also the center of a circle inscribed in that same polygon.

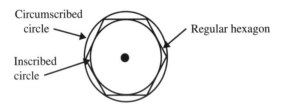

• A *circle is circumscribed about a polygon*, or a *polygon is inscribed in a circle*, if and only if each vertex of the polygon lies on the circle.

Square   Hexagon   Octagon

Each vertex of the polygon lies on the circle.

• A circle is *inscribed in a polygon*, or a polygon is *circumscribed about a circle*, if and only if *each side of the polygon is **tangent*** to the circle.

Each side of the polygon is *tangent* to the circle.

- *Triangle incircle theorem*: Every triangle has an *incircle*.

We can show this by drawing radii from the center O of the circle to each side of the triangle. The radii are congruent, $OD \cong OE \cong OF$, and perpendicular to a line at the point of tangency, $OD \perp AB$, $OE \perp BC$, and $OF \perp AC$. Because each side of the polygon is *tangent* to the circle, it is an incircle (inscribed in) of the polygon, in this case a triangle.

- If a polygon is cyclic, then a circle exists that contains all of its vertices.

- *Cyclic quadrilateral theorem*: A quadrilateral is cyclic if and only if its opposite angles are supplementary.

- *Converse of cyclic quadrilateral theorem*: In a quadrilateral if a pair of its opposite angles are supplementary, then it is cyclic.

- **Example:** Prove informally that if a quadrilateral is cyclic, then its opposite angles are supplementary.

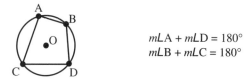

$$mLA + mLD = 180°$$
$$mLB + mLC = 180°$$

Given: Quadrilateral ABDC is cyclic and circle O contains its vertices.

Prove: LB and LC are supplementary.

Proof: Because an inscribed angle is equal to half its intercepted arc:

$mLB = (1/2) \, m$ arcACD and $mLC = (1/2) \, m$ arcABD.

By addition: $mLB + mLC = (1/2) \, m$ arcACD $+ (1/2) \, m$ arcABD $= (1/2)(m$ arcACD $+ m$ arcABD$)$

Because arcACD and arcABD combine to make up the circle, which measures 360°, then m arcACD $+ m$ arcABD $= 360°$. Therefore:

$$mLB + mLC = (1/2)(m \text{ arcACD} + m \text{ arcABD}) = (1/2)(360°) = 180°$$

Because $mLB + mLC = 180°$, then LB and LC are supplementary.

6.8 Chapter 6 Summary and Highlights

• A *circle* is a planar shape consisting of a closed curve in which each point on the curve is the same distance from the center of the circle. The *radius* of a circle is the distance between the center and any point on the circle. A line segment drawn through the center point with its endpoints on the circle is a *diameter* of the circle. A *chord* is a line segment that joins two points of a circle. A *secant* is a line that intersects a circle in two points and contains a chord. *Congruent circles* are circles that have congruent radii and congruent diameters. *Concentric circles* lie in the same plane and have the same center point.

• Any angle whose vertex is at a circle's center point with its sides as radii is called a *central angle*, and any angle whose vertex is on the circle with its sides as chords is called an *inscribed angle*.

If an *inscribed angle* intercepts a semicircle, it is a right angle. Inscribed angles intercepting the same arc or congruent arcs are equal in measure.

• An *arc* is a continuous section of a circle between two points and includes the endpoints. There are three types of arcs: *semicircles* (half the circumference), *minor arcs* (less than half the circumference), and *major arcs* (greater than half the circumference). The *measure of a semicircle* is 180° or π. The *measure of a minor arc* in degrees is the measure of its corresponding central angle. The *measure of a major arc* in degrees is: (360° – measure of its minor arc in degrees). An arc can be measured by its *central angle* or by its *length as a fraction of the circumference*. The *length of an arc* equals the fraction of the circle it covers ($m°/360°$) multiplied by the circumference ($2\pi r$): Arc length = ($m°/360°$)$2\pi r$, or (arc length)/(circumference $2\pi r$) = ($m°$)/(360°).

• *Chords*: In the same circle or in congruent circles, congruent chords have congruent arcs, and congruent arcs have congruent chords. Chords of equal distance from the center of a circle are congruent. *Perpendicular bisectors of chords* in a circle pass through its center. The measure of an angle formed by a *chord and a tangent* is equal to one-half the measure of the intercepted arc. If two chords intersect inside a circle, the product of the lengths of the segments of one chord is equal to the product of the lengths of the segments of the other chord.

- *Secants and tangents*: A *secant angle* whose vertex is inside a circle is equal in measure to half the sum of the arcs intercepted by it and its vertical angle. A *secant angle* whose vertex is outside a circle is equal in measure to one-half the difference between the measures of the arcs intercepted by it. If *two secant segments* are drawn to a circle from a point outside the circle, the product of the lengths of one (entire) secant segment and its external segment is equal to the product of the lengths of the other (entire) secant segment and its external segment. If a *secant segment and a tangent segment* are drawn to a circle from an external point, the product of the lengths of the (entire) secant segment and its external segment is equal to the square of the tangent segment. If a line is *tangent* to a circle, the line is *perpendicular* to the radius. If a tangent line meets a diameter, they are *perpendicular* to each other. *Tangent segments* to a circle from an external point are congruent. A *common tangent* is a line that is tangent to two coplanar circles.

- *Circumference* of a circle $= 2\pi r = \pi d$. *Area* of a circle $= \pi r^2 = \pi(d/2)^2$.

- Every *regular polygon* is cyclic.

- The *area of a sector* whose arc has measure $m°$ is:

    Area of a sector $= (m°/360°)\pi r^2$, or

    (area of a sector)/(area of a circle πr^2) $= (m°)/(360°)$.

Chapter

7

Geometric Solids: Surface Area and Volume of Three-Dimensional Objects

Two-dimensional Euclidean geometry is called *plane geometry* and studies planar figures that can be constructed using a straightedge and compass. Three-dimensional Euclidean geometry is called *solid geometry*. Most of the objects we look at or handle in everyday life are three-dimensional solids. In addition, most of us live in some variation of a rectangular solid, we engage in or watch sports that are centered around spheres or spheroids, we use liquids that are stored in three-dimensional containers, we use water that is delivered to us in cylindrical pipes, we are intrigued by the great pyramids in Egypt, and we look to the heavens and see spherical bodies and spiral and elliptical galaxies. Even molecules exist in nature in geometric configurations such as pyramidal, tetrahedral, and octahedral shapes. Molecules build familiar compounds such as NaCl, which forms a cubic lattice, and quartz (SiO_2), which occurs as a hexagonal crystal. The world of three-dimensional objects includes prisms such as cubes and rectangular solids, as well as cylinders, pyramids, cones, spheres, and numerous multi-sided solids such as tetrahedrons, octahedrons, and dodecahedrons.

7.1 Solids

• Three-dimensional objects, or solids, take up space in three dimensions as opposed to the two-dimensional figures we have studied so far in this book, which reside in planes. Measurements of three-dimensional objects include volume and surface area, as well as diagonal lengths. *Volume* is a measure of the three-dimensional space that an object occupies. The units for volume are always cubed because of the three dimensions described, $(x)(x)(x) = x^3$. The *surface area* of three-dimensional objects is a sum of the areas of the surfaces. The units for surface area are always squared because of the two dimensions described, $(x)(x) = x^2$. Remember to convert all measurements to the same units (such as inches or meters) before calculating.

• Two- or three-dimensional objects can have their locations changed and be unchanged themselves. Describing this is the *change position postulate*, which states that any geometric figure can be moved or relocated to a new position without changing the figure's size or shape.

• A *polyhedron* is a three-dimensional solid figure that is bounded by intersecting planes. The intersection of planes forms polygonal-shaped surfaces, which are called the *faces* of the polyhedron. The edges of the polygonal-shaped faces formed at the intersection of planes are called *edges*, and the vertices of the polygonal faces are the *vertices* of the polyhedron. A polyhedron can be defined as a three-dimensional solid consisting of an assembly of polygons usually joined at their edges. The word *polyhedron* is derived from the Greek *poly* (many) plus the Indo-European *hedron* (seat). Polyhedrons include prisms; pyramids; the *Platonic solids*, including tetrahedrons, cubes, octahedrons, dodecahedrons, and icosahedrons; the Archimedean solids, such as cuboctahedrons and icosidodecahedrons; and the Johnson solids, such as square pyramids and triangular cupolas (dome-shape). The five Platonic solids that Plato used to explain the structure of matter are convex regular polyhedra having faces that are identical regular polygons. The five Platonic solids are:

Tetrahedron   Cube   Octahedron   Dodecahedron Icosahedron

Tetrahedron: 4 triangular faces; 3 triangles meet at each vertex.

Hexahedron or cube: 6 square faces; 3 squares can meet at a vertex.

Octahedron: 8 triangular faces; 4 triangles meet at each vertex.

Dodecahedron: 12 pentagonal faces; 3 pentagons meet at a vertex.

Icosahedron: 20 triangular faces; 5 triangles meet at each vertex.

• **Example:** If you plan to construct a tetrahedron-shape glass figure with an edge length of 2 meters, what is the surface area of glass needed?

We need to find the surface area of a regular tetrahedron with each edge e having a length of 2 meters. The surface area of a tetrahedron is the sum of its four faces, and in the case of a regular tetrahedron, the four faces are each congruent equilateral triangles with congruent areas, so that: $\triangle ABC \cong \triangle ADC \cong \triangle BDC \cong \triangle ABD$.

In Section 4.13. *Area of a Triangle* we derived from *Heron's Theorem* that the *area of an equilateral triangle* is: $(1/4)(\text{side})^2 \sqrt{3}$

The total surface area of the tetrahedron is therefore:

$$(4)(\text{area } \triangle ABC) = (4)(1/4)(e^2)\sqrt{3} = (e^2)\sqrt{3}$$

If $e = 2$ meters, then the surface area is $4\sqrt{3}$ meters$^2 \approx 6.928$ square meters and we need this amount of glass.

7.2 Prisms: Cubes, Rectangular Solids, and Oblique and Right Prisms

• *Prisms* are three-dimensional objects, or solids, having two congruent polygonal bases that lie in parallel planes. Examples of prisms include cubes having six surfaces that are squares, rectangular solids having six surfaces that are rectangles or parallelograms, triangular prisms having two triangular bases and three lateral surfaces that are rectangles or parallelograms, and pentagonal prisms having two pentagonal bases and five lateral sides that are rectangles or parallelograms.

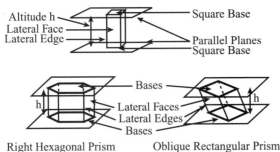

Right Hexagonal Prism   Oblique Rectangular Prism

• **Prisms possess the following**:

Two bases that are ***congruent polygons*** and lie in parallel planes.

Lateral faces that connect the two congruent bases and are parallelograms or rectangles (for right prisms).

Lateral edges, which are parallel line segments connecting the corresponding vertices of the two bases.

An ***altitude*** joining the two base planes with its length as the ***height*** h of the prism. The altitude segment is perpendicular to the planes containing the bases, has its endpoints on each plane, and measures the *distance* between the two bases.

• ***Right prisms*** and ***oblique prisms***:

A ***right prism*** is a prism whose *lateral edges are perpendicular to the two bases*. In a right prism, the altitude lies on, or is parallel to, a lateral edge.

An ***oblique prism*** is a prism whose *lateral edges are not perpendicular to the two bases*. In an oblique prism, the altitude does not lie on, or is not parallel to, a lateral edge.

• *Prisms are classified* by both the shape of their bases and the relation (either right or oblique) of their lateral edges to the planes containing their bases. The preceding figure depicts a *right hexagonal* prism and an *oblique rectangular* prism.

Surface Area

• The ***surface area of a prism*** is the sum of the surface areas of its sides. It is generally easier to separately calculate the area of the two congruent bases, 2B, and the area of the lateral faces, or lateral area LA.

Total area TA = lateral area LA + (2) base area B

TA = LA + 2B

- If a prism is a **right prism**:

Right prism lateral area theorem: The lateral area LA of the lateral faces of a right prism having altitude h and perimeter of a base p is:

$$LA_{\text{right prism}} = (p_{\text{base}})(h) \text{ units}^2$$

Right prism total area theorem: The surface area TA of a right prism is equal to the area of the two congruent bases, 2B, plus the lateral area LA, which is the perimeter of a base p multiplied by its height h.

$$TA_{\text{right prism}} = 2B + LA_{\text{right prism}} = 2B + p_{\text{base}}h_{\text{prism}} \text{ units}^2$$

B is the area of base.
p is the perimeter of base.
h is the height of prism.

Calculating the areas of polygons is discussed in Chapter 5, Section 5.8. *Area and Perimeter.*

- **Example:** Determine the surface area of the trapezoidal right prism.

4 Ft
6 Ft 4 Ft ——— Altitudes of trapezoid bases are 4 feet.
3 Ft ———— Height of prism is 3 feet.
10 Ft

First, find the perimeter and area of one isosceles trapezoidal base.

$$\text{Perimeter}_{\text{(trapezoid)}} = \text{side } 1 + \text{side } 2 + \text{side } 3 + \text{side } 4$$
$$= 4 \text{ feet} + 6 \text{ feet} + 6 \text{ feet} + 10 \text{ feet} = 26 \text{ feet}$$

$$\text{Area}_{\text{(trapezoid)}} = (\text{average of trapezoid's bases})(\text{altitude}_{\text{trapezoid}})$$
$$= (1/2)(\text{base } 1_{\text{trap}} + \text{base } 2_{\text{trap}})(\text{altitude}_{\text{trapezoid}})$$
$$= (1/2)(4 \text{ feet} + 10 \text{ feet})(4 \text{ feet}) = 28 \text{ square feet}$$

Therefore, the perimeter is 26 feet and the area is 28 square feet for the trapezoidal bases of the prism.

Next, calculate the lateral area of the prism: $LA_{\text{right prism}} = (p_{\text{base}})(h_{\text{prism}})$

$$LA_{\text{right prism}} = (26 \text{ ft.})(3 \text{ ft.}) = 78 \text{ square feet}$$

Finally, find the total area of the prism:

$\text{TA}_{\text{right prism}} = (2)(\text{base area}) + \text{LA}_{\text{right prism}}$

$\text{TA}_{\text{right prism}} = (2)(28 \text{ square feet}) + (78 \text{ square feet})$

$\text{TA}_{\text{right prism}} = 134 \text{ square feet}$

The total surface area of the trapezoidal right prism is 134 square feet.

Volume

• *Volume* measures the interior three-dimensional space of a three-dimensional object. The dimensions can be thought of as containing one-by-one unit cubes. A cube is a square right prism having six congruent square surfaces. The interior volume of any right rectangular prism can be filled with unit cubes resulting in the solid having a measurement of some number of cubic units, such as cubic inches.

 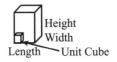

• The volume of a rectangular solid is:

$V = \text{length} \times \text{width} \times \text{height}$

$\quad = \quad \text{base} \quad \times \quad \text{height}$

Where the area of the base is: length × width.

• *Prism volume postulate*: The volume V of any prism is Bh, where B is the area of one of its bases and h is the length of the prism's altitude.

$$V_{\text{prism}} = \text{Bh units}^3$$

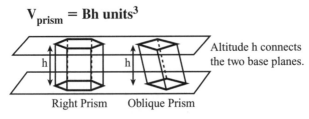

Altitude h connects the two base planes.

Right Prism   Oblique Prism

Note: The bases of a prism are congruent. If the area of the bases of two prisms are congruent, and their heights are also congruent, then the volumes are congruent, even if their shapes are different. See Section 7.8. *Cavalieri's Prinicple*.

• **Example:** If you have a wading pool shaped like the trapezoidal right prism in the preceding example, determine the volume of water needed to fill the pool.

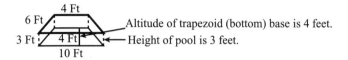

Altitude of trapezoid (bottom) base is 4 feet.

Height of pool is 3 feet.

The volume of the prism is $V = Bh$, where B is the area of the (bottom) base of the prism and h is the height of the prism.

In the preceding example, we determined the area B of the trapezoid base:

$$B = Area_{trapezoid} = (1/2)(base\ 1_{trapezoid} + base\ 2_{trapezoid})(altitude_{trapezoid})$$
$$= (1/2)(4\ feet + 10\ feet)(4\ feet) = 28\ square\ feet$$

$$h_{prism} = 3\ feet$$

$$V_{prism} = Bh = (28\ square\ feet)(3\ feet) = 84\ feet^3\ or\ 84\ cubic\ feet$$

We need 84 cubic feet of water to fill the wading pool.

• *Cubes* have six surfaces that are each squares and have the same measurements. A cube is a square right prism and its lateral edges are equal in length to the sides of the base. If the cube below has a side length of 5 inches, we can calculate its volume, surface area, and main diagonal.

5 in

The volume of a prism is Bh.

For a cube, B = length × width = edge × edge, and h = edge, so that:

Volume of a cube $= (edge)^3 = (5\ inches)^3 = 125\ inches^3$

To determine surface area, we can calculate the area of a face and multiply by the number of faces, 6, (the bases and lateral faces are equal), or we can determine surface area using the right prism area theorem.

Surface area of a cube = (6 sides)(area of each square side)

$$= (6\ sides)(5\ inches)^2 = (6\ sides)(25\ inches^2) = 150\ inches^2$$

or,

$$TA_{right\ prism} = (2)(base\ area) + LA_{right\ prism} = 2B + p_{base}h_{prism}$$

where perimeter $p = 5 × 4 = 20$, h = 5, and base area $B = 5 × 5 = 25$.

Surface area $= TA_{right\ prism} = (2)(25) + (20)(5) = 150\ square\ inches$

The *main diagonal* d of a cube is given by: $d^2 = l^2 + w^2 + h^2$, or

$d^2 = e^2 + e^2 + e^2 = 3e^2$, where e = the length of each edge.

$d^2 = 3e^2 = 3(5)^2 = 75$, or $d^2 = 75$

Take the square root of the equation: $d \approx 8.7$ inches

Therefore, the volume is 125 cubic inches, the surface area is 150 square inches, and the main diagonal is approximately 8.7 inches.

• **Example:** If the total surface area of a cube is 150 square centimeters, what is the length of an edge? After determining the edge, find the volume.

We can use: $TA_{right\ prism} = 2B + LA_{right\ prism} = 2B + p_{base}h$

In our cube: h = e, p = 4e, and $B = (e)(e) = e^2$.

Substitute into the total area equation:

$TA_{right\ prism} = 2(e)^2 + (4e)(e) = 2e^2 + 4e^2 = 6e^2$

In this example TA = 150 cm^2, so we can substitute and solve for e:

$150\ cm^2 = 6e^2$

$e^2 = 150\ cm^2/6 = 25\ cm^2$

Take the square root:

e = 5 centimeters.

The volume is $V_{prism} = Bh$

For a cube, $B = e \times e$ and h = e, so volume is e^3, or:

$V = (5\ centimeters)^3 = 125$ cubic centimeters

Therefore, the edge is 5 centimeters and the volume is 125 cubic centimeters.

• *Rectangular solids* have six rectangular surfaces with three pairs of opposite surfaces that have the same measurements. The rectangular solid below has a length of 8 inches, a width of 2 inches, and a height of 3 inches. We can determine its volume, surface area, and main diagonal as follows:

Width = 2 in.
Height = 3 in.

Length = 8 in.

Volume of a rectangular solid = Bh = (length)(width)(height)

= (8 in.)(2 in.)(3 in.) = 48 inches^3 or 48 cubic inches

Surface area of a rectangular solid = the sum of the areas of its 6 faces, where opposite sides are identical.

Area = (2)(length)(width) + (2)(length)(height) + (2)(width)(height)

= (2)(8)(2) + (2)(8)(3) + (2)(2)(3) = 32 in.^{2} + 48 in.^{2} + 12 in.^{2} = 92 in.^{2}

We can also use the right prism area theorem:

$TA_{right\ prism}$ = (2)(base area) + $LA_{right\ prism}$ = 2B + $p_{base}h_{prism}$

where perimeter p = 8 + 8 + 2 + 2 = 20, h = 3, and B = 8 × 2 = 16.

Surface area = $TA_{right\ prism}$ = (2)(16) + (20)(3) = 92 square inches

The *main diagonal* of a rectangular solid is given by: $d^2 = l^2 + w^2 + h^2$.

d^2 = (8 in.)^2 + (2 in.)^2 + (3 in.)^2 = 64 in.^{2} + 4 in.^{2} + 9 in.^{2} = 77 in.^{2}

Take the square root: d ≈ 8.8 inches

Therefore, the volume is 48 cubic inches, the surface area is 92 square inches, and the main diagonal is approximately 8.8 inches.

7.3 Pyramids

• The largest constructed geometric solid is the Great Pyramid in Egypt, which has a base of approximately 13 acres and an original height of 481 feet. *Pyramids* are three-dimensional objects that have a square, rectangle, triangle, or other polygon base in a plane connected to a point outside the plane. The set of all segments joining the point to the polygon base on the plane forms the pyramid. In ***regular pyramids***, the bases are regular polygons (having congruent sides and congruent angles) and their lateral edges are congruent.

Regular triangular
pyramid

Regular square
pyramid

Regular pentagonal
pyramid

Oblique
pyramid

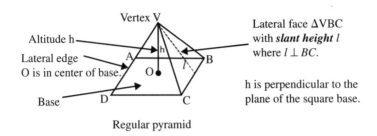

Vertex V

Altitude h

Lateral edge

O is in center of base.

Base   D   C

Lateral face ΔVBC
with **slant height** *l*
where *l* ⊥ *BC*.

h is perpendicular to the
plane of the square base.

Regular pyramid

The above pyramid has a square base, ABCD, and four triangular lateral faces, ΔVAD, ΔVDC, ΔVBC, and ΔVAB. In a *regular pyramid* there is a measure called the *slant height, l*, which is a segment from the vertex to the base of each side.

- **Pyramids possess the following**:

  One *polygon base* in a plane and a vertex point not on the base plane.

  Triangular *lateral faces* that connect the base to the vertex point.

  Lateral edges formed by intersecting faces. Lateral edges are line segments connecting the vertices of the base to the vertex point.

  An *altitude* h, which is a segment extending from the vertex point perpendicular to the plane of the base. Its length h is the **height** *of the pyramid.*

- If a pyramid is a *regular pyramid*:

  The *base* is a regular polygon.

  All *lateral edges* are congruent.

  All *lateral faces* are *congruent isosceles triangles.*

  The lateral faces have a *slant height l*, which extends from the vertex perpendicular to the side of each base (not perpendicular to the base plane).

  The *altitude* h extends from the vertex perpendicular to the center of the polygon base.

Surface Area

- The **surface area** of a pyramid is the surface area of its base, plus the area of its lateral faces.

- The **lateral area of a regular pyramid** can be determined by either of the following:

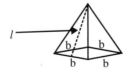

Base perimeter p = b + b + b + b

(1) Calculate the area of one lateral face (isosceles triangle) and multiply by the number of faces:

The area of one lateral face is $(1/2)$(base of Δ face)(slant height l)

The lateral area of the pyramid is:

$$LA = (1/2)\text{(base of triangle face b)(slant height } l)\text{(n sides)}$$
$$= (1/2)nbl$$

(2) Use the *regular pyramid lateral area theorem*:

$$LA_{\text{regular pyramid}} = (1/2)pl$$

Regular pyramid lateral area theorem: The lateral area LA of the lateral faces of a regular pyramid having slant height l and perimeter p of the base is:

$$LA_{\text{regular pyramid}} = (1/2)pl \text{ units}^2$$

• The *total surface area* of a regular pyramid can be found using the **regular pyramid total area theorem**: The *surface area* TA of a regular pyramid is equal to the lateral area LA, plus the area B of the base.

$$TA_{\text{regular pyramid}} = B + LA_{\text{regular pyramid}}$$
$$= B + (1/2)pl \quad \textbf{units}^2$$
$$= B + (1/2)nbl \quad \textbf{units}^2$$

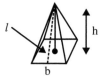

B = area of base
p = perimeter of base
l = slant height of face
n = number of faces
b = base of triangle face

Regular pyramid

The area formulas for various polygonal bases are discussed in Chapter 5, Section 5.8. *Area and Perimeter of Squares, Rectangles, Parallelograms, Rhombuses, Trapezoids, Other Polygons, and Regular Polygons.*

Volume

• *Pyramid volume theorem*: The volume V of any pyramid with base area B and altitude h is given by:

$$V_{\text{pyramid}} = (1/3)Bh \text{ units}^3$$

B = area of base
h = height of pyramid

• *Side note*: If a pyramid and a prism have congruent bases and equal heights, the volume of the pyramid will be one-third the volume of the prism.

$V_{prism} = Bh$ $V_{pyramid} = (1/3)Bh$

• When you are solving problems involving ***regular pyramids***, note that within the pyramid, right triangles exist and their properties can be used to determine unknown lengths. In the figure below:

Each slant height l, lateral edge, and half of the base forms a right triangle, such as $\triangle VSB$ and $\triangle VSC$. Unknown lengths can be determined using the Pythagorean Theorem, $leg^2 + leg^2 = hypotenuse^2$.

Each segment from the center of the base O to a slant height, the slant height segment l, and pyramid height h forms a right triangle, such as $\triangle VOS$. Unknown lengths can be determined using the Pythagorean Theorem, $leg^2 + leg^2 = hypotenuse^2$.

Each side has right triangles, with $\triangle VSB$, $\triangle VSC$, and $\triangle VOS$ denoted for side BCV.

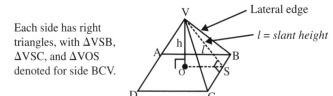

Lateral edge

l = slant height

• **Example:** You want to construct a small model pyramid with a square base in your backyard (sketched below). If a side of the square base is 10 meters and the height of the pyramid is 8 meters, what will be the volume and surface area?

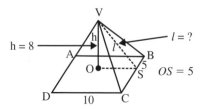

$h = 8$ $l = ?$ $OS = 5$

The surface area and volume can be calculated using:

$TA_{regular\ pyramid}$ = (base area B) + $LA_{regular\ pyramid}$ = B + (1/2)pl

$V_{pyramid} = (1/3)Bh$

The area of the square base B $= 10^2 = 100$ meters^2.

The perimeter of the base p $= 4 \times 10 = 40$ meters.

The pyramid height h $= 8$ meters.

To obtain the lateral area, the lateral slant height l must be determined. Because the lateral slant height is part of a right triangle ΔVOS in which the long leg h $= 8$ is perpendicular to the base, the short leg is half the base, or 5, and the slant height l is the hypotenuse. We can use the Pythagorean Theorem: $leg^2 + leg^2 = hypotenuse^2 \rightarrow 8^2 + 5^2 = l^2$

$l^2 = 64 + 25 = 89$ meters^2

The square root of $89 \approx 9.43$ meters

Therefore, the slant height l is approximately 9.43 meters.

$TA_{regular\ pyramid} = B + (1/2)pl$ units^2

$= (100$ meters$^2) + (1/2)(40$ meters$)(9.43$ meters$) \approx 288.6$ meters^2

$V_{pyramid} = (1/3)Bh = (1/3)(100$ meters$^2)(8$ meters$) \approx 266.7$ meters^3

Therefore, your pyramid will have a volume of approximately 266.7 cubic meters and a surface area of approximately 288.6 square meters.

7.4 Cylinders

• *Cylinders,* or *circular solids,* are three-dimensional objects that have two identical circle bases in parallel planes connected by the set of all line segments between the two circle bases. If the segments joining the centers of the circles are perpendicular to the two planes of the circles, the cylinder is a *right circular cylinder*. If the segments joining the centers of the circles are *not* perpendicular to the planes of the circles, the cylinder is an *oblique circular cylinder*.

Right circular cylinder   Oblique circular cylinder

• **Cylinders possess the following**:

Two congruent *circle bases* in parallel planes.

Lateral surface that is curved and connects the two bases.

An *altitude* h, which is a segment joining the two base planes that is perpendicular to the planes. Its length h is the *height of the cylinder*.

• If a cylinder is a *right cylinder*, the *altitude* h joins the centers of the bases at right angles.

• Cylinders are similar to prisms, except the bases are circles, so that determining surface area and volume is similar.

Surface Area

• The ***total surface area*** of a circular cylinder is the area of the two congruent circle bases plus the lateral area (surface connecting the bases).

• ***Right circular cylinder lateral area theorem***: The lateral surface area LA of a right circular cylinder having an altitude h and the perimeter of its circle bases as the *circumference* $C = 2\pi r$ is:

$LA_{\text{right cylinder}} = Ch = 2\pi rh$ units^2

• ***Right circular cylinder total area theorem***: The surface area TA of a right circular cylinder is equal to the lateral area, $LA = Ch$, plus the area B of the two congruent circle bases. (The *area of a circle is* πr^2.)

$$TA_{\text{right cylinder}} = 2B + LA_{\text{right cylinder}} = 2B + Ch$$
$$= 2\pi r^2 + 2\pi rh \text{ units}^2$$

h = height
C = 2πr
πr² = circle base area

Volume

• ***Cylinder volume theorem***: The volume V of any circular cylinder having two congruent circle bases with area $B = \pi r^2$ and altitude h is:

$V_{\text{cylinder}} = Bh = \pi r^2 h$ units^3

• **Example:** In the cylinder below, if r = 2 inches and h = 10 inches, what is the volume and surface area?

Volume of a cylinder = (area of circle base B)(height h) = $(\pi r^2)(h)$
$= (\pi)(2 \text{ in})^2(10 \text{ in}) = 40\pi \text{ in}^3 \approx 125.7$ inches^3 or 125.7 cubic inches.

Surface area of a right circular cylinder

= (area of both circle bases) + (lateral area Ch)

$= 2\pi r^2 + 2\pi rh$

$= (2)(\pi)(2 \text{ in.})^2 + (2)(\pi)(2 \text{ in.})(10 \text{ in.}) = 8\pi \text{ in.}^2 + 40\pi \text{ in.}^2 = 48\pi \text{ in.}^2$

$\approx 150.8 \text{ in.}^2$ or 150.8 square inches

Therefore, the volume is approximately 125.7 cubic inches, and the surface area is approximately 150.8 square inches.

7.5 **Cones**

• *Cones* are three-dimensional objects that have a circle base connected to a vertex point outside the plane of the base. The depth of a cone forms a triangular solid. When the segment joining the center of the circle base and vertex point is perpendicular to the base, the cone is a *right circular cone*. When the segment joining the center of the circle base and vertex point is *not* perpendicular to the base, the cone is an *oblique circular cone*.

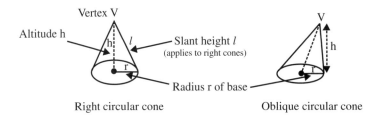

Right circular cone   Oblique circular cone

• *Cones possess the following*:

One *circle base* in a plane and a vertex point not on the base plane.

Lateral face that connects the base to the vertex point.

An *altitude* h, which is a segment extending from the vertex point perpendicular to the plane of the base. Its length h is the *height* of the cone.

• If a cone is a *right circular cone*:

The *altitude* h extends from the vertex perpendicular to the center of the circle base.

The lateral surface has a *slant height* l, which extends from the vertex to the base along the lateral surface.

Also, if the lateral area of a right cone is cut along l and flattened out, it has the shape of a sector of a circle.

• Cones are similar to pyramids, except the bases are circles, so that determining surface area and volume is similar.

Surface Area

• The *total surface area* of a cone is the area of the circle base and the lateral area between the base and the vertex point.

• *Right circular cone lateral area theorem*: The lateral surface area LA of a right circular cone having a *slant height l* and a perimeter of its base that is the *circumference* $C = 2\pi r$ is:

$$LA_{\text{right cone}} = (1/2)Cl = (1/2)2\pi rl$$
$$= \pi rl \text{ units}^2$$

• *Right circular cone total area theorem*: The surface area TA of a right circular cone is equal to the lateral area, $LA = \pi rl$, plus the circle *base area* B, which is the area of a circle $= \pi r^2$.

$$TA_{\text{right cone}} = B + LA_{\text{right circular cone}}$$
$$= \pi r^2 + \pi rl \text{ units}^2$$

l = slant height
πr^2 = area of circle base
πrl = lateral area

Volume

• *Cone volume theorem*: The volume V of a circular cone having a circle base with area $B = \pi r^2$ and altitude h is given by:

$$V_{\text{cone}} = (1/3)Bh = (1/3)\pi r^2 h \text{ units}^3$$

• *Side note*: If a cone and a cylinder have congruent radii and heights, the volume of the cone will be one-third the volume of the cylinder.

$V_{\text{cylinder}} = Bh = \pi r^2 h$ $V_{\text{cone}} = (1/3)Bh = (1/3)\pi r^2 h$

• **Example:** You want to construct a small model right cone in your backyard (sketched below). If the radius of the circle base is 4 meters and the height of the cone is 8 meters, what will be its volume and surface area?

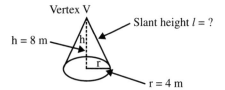

Vertex V

Slant height l = ?

h = 8 m

h

r

r = 4 m

The surface area and volume can be calculated using:

$$TA_{right\ cone} = B + LA_{right\ cone} = \pi r^2 + \pi r l$$
$$V_{cone} = (1/3)Bh = (1/3)\pi r^2 h$$

The area of the circle base $B = \pi r^2 = \pi 4^2 = 16\pi$ meters^2

The cone height h = 8 meters.

To determine the lateral area, first the lateral slant height l must be calculated. Because the lateral slant height is part of a right triangle with long leg h perpendicular to short leg r, we can use the Pythagorean Theorem: $\text{leg}^2 + \text{leg}^2 = \text{hypotenuse}^2 \rightarrow 4^2 + 8^2 = l^2$

$$l^2 = 16 + 64 = 80$$

The square root of $80 = l \approx 8.94$ meters

$$TA_{right\ cone} = \pi r^2 + \pi r l = \pi 4^2 + \pi(4)(8.94)$$
$$= (16\pi\ \text{meters}^2) + (\pi)(4\ \text{meters})(8.94\ \text{meters}) = 51.76\pi\ \text{meters}^2$$
$$\approx 162.6\ \text{meters}^2$$

$$V_{cone} = (1/3)\pi r^2 h = (1/3)(16\pi\ \text{meters}^2)(8\ \text{meters}) = 42.67\pi\ \text{meters}^3$$
$$\approx 134\ \text{meters}^3$$

Therefore, the surface area is approximately 162.6 square meters and the volume is approximately 134 cubic meters.

7.6 Spheres

• If a slice is made through the center of a sphere, and the sphere is split into two equal hemispheres, the diameter of the circle formed by the slice will be the same as the diameter of the sphere. The circle formed is called a **great circle**, and is the largest circle that can be drawn on the surface of a sphere. The *equator* on Earth is a *great circle* and is halfway between the North and South Poles. (The Earth is not a perfect sphere, but is often modeled as a sphere.) When measuring distances or plotting a course on a sphere, it turns out that the *shortest distance or path between two points is the path that is part of a great circle* on which the two points are located. In other words, on a spherical surface, a great circle path, often called a **geodesic**, is always the shortest path between two points.

• An example of using the properties of a sphere is the *geodesic dome* patented by R. Buckminster Fuller. It is a framework constructed with straight pieces of steel or aluminum tubing in a network of triangles that is covered with thin aluminum or plastic. While the segments forming the network can be different lengths, the vertices are all equidistant from the center of the dome and lie on a sphere. Chains of segments around the dome form a circular geodesic path so that the dome, in essence, uses a network of great circles within the surface of a sphere as supports to distribute stress, making it strong. It is known for its strength and wind resistance.

• Another example of the use of spherical shapes in architecture is the triangular spherical shells of the famous Sidney Opera House in Sidney, Australia.

• Another interesting property of a sphere is that it holds a greater volume than any other container with the same surface area.

• *Spheres* or *spherical solids* are three-dimensional objects consisting of points that are all the same distance from a center. A sphere is the set of all points in three-dimensional space that are equal distant (the radius) from a given point, which is the sphere's center.

• *Sphere surface area theorem*: The surface area of a sphere with radius r is:

$$A_{sphere} = 4\pi r^2 \text{ units}^2$$

• *Sphere volume theorem*: The volume of a sphere with radius r is:

$$V_{sphere} = (4/3)\pi r^3 \text{ units}^3$$

• Note that the radius is the only variable used to determine surface area and volume of a sphere.

• **Example:** If the radius of a sphere is 2 feet, determine the volume and surface area.

   r = 2 feet

The volume is: $V_{sphere} = (4/3)\pi r^3$

$V_{sphere} = (4/3)(\pi)(2 \text{ ft})^3 = (4/3)(\pi)(8 \text{ ft}^3) = (32/3)\pi \text{ ft}^3 \approx 33.5 \text{ feet}^3$

The surface area of a sphere is: $A_{sphere} = 4\pi r^2$

$A_{sphere} = 4\pi(2 \text{ ft})^2 = 4\pi(4 \text{ ft}^2) = 16\pi \text{ ft}^2 \approx 50.26 \text{ feet}^2$

Therefore, the volume is approximately 33.5 cubic feet and the surface area is approximately 50.26 square feet.

• **Example:** You are that the average circumference of the Earth is approximately 24,900 miles. What are the approximate values of its surface area and volume?

$C = 24,900$ miles

We can approximate the Earth as a sphere and first determine its radius using $C = 2\pi r$, or $r = C/2\pi = 24,900 \text{ miles}/2\pi = 12,450/\pi \approx 3,963$ miles

Using the radius, we can determine surface area and volume:

$A_{sphere} = 4\pi r^2 = 4\pi(3,963 \text{ miles})^2 \approx 197,359,488$ square miles

$V_{sphere} = (4/3)\pi r^3 = (4/3)\pi(3,963 \text{ miles})^3 \approx 260,711,800,000$ cubic miles

The approximate values of Earth's surface area and volume are 197,359,488 square miles and 260,711,800,000 cubic miles, respectively.

7.7 Similar Solids

• Similar solids, like similar polygons, have the same shape, but not necessarily the same size. If two solids are similar, the bases are similar and the corresponding lengths are in proportion.

• *Similar solids surface area theorem*: The ratio of the surface areas of two similar solids is equal to the *square of the ratio* of any pair of corresponding lengths, so that if the ratio of lengths (scale factor) is

   x : y, then the ratio of areas is $x^2 : y^2$.

• *Similar solids volume theorem*: The ratio of the volumes of two similar solids is equal to the *cube of the ratio* of any pair of corresponding lengths, so that if the ratio of lengths (scale factor) is

   x : y, then the ratio of volumes is $x^3 : y^3$.

• In the following similar solids, the bases are similar and the corresponding lengths are in proportion. We will see in the following example that if corresponding lengths have a ratio of x : y, then the ratio of corresponding areas is $x^2 : y^2$, and the ratio of corresponding volumes is $x^3 : y^3$.

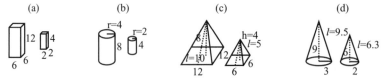

(a)   (b)   (c)   (d)

B is 6 × 6 and 2 × 2;   r is 4 and 2;   B is 12 × 12 and 6 × 6;   r is 3 and 2; h is 9 and 6;
h is 12 and 4;   h is 8 and 4;   h is 8 and 4; *l* is 10 and 5;   *l* is 9.5 and 6.3;
6/2 = 12/4 = 3/1.   4/2 = 8/4 = 2/1.   12/6 = 10/5 = 8/4 = 2/1.   9/6 = 3/2 = 9.5/6.3 = 1.5/1.

In the four pairs of similar figures, corresponding lengths are in proportion so that: side lengths/side lengths = heights/heights = radii/radii = slant heights/slant heights.

• **Example:** Show the ratios of perimeter or circumference lengths x : y, and verify that the ratio of corresponding surface areas is $x^2 : y^2$, and the ratio of corresponding volumes is $x^3 : y^3$ for the similar pairs of solids in the preceding figure: (a) rectangular solids, (b) right cylinders, (c) regular pyramids, and (d) right cones.

(a)   Rectangular solids:

Perimeter ratio: 24/8 = 3:1

Area is: $TA_{right\ prism} = 2B + p_{base}h$ units^2

Ratio $TA_{right\ prism} / TA_{right\ prism}$ is: $[(2)(36) + (24)(12)] / [(2)(4) + (8)(4)]$
$$= 360 / 40 = 9:1 = 3^2:1^2$$

Volume is: $V_{prism} = Bh$ units^3

Ratio V_{prism} / V_{prism} is: $[(36)(12)] / [(4)(4)] = 432 / 16 = 27:1 = 3^3:1^3$

(b)   Right cylinders:

Circumference ($2\pi r$) ratio: $8\pi/4\pi = 2:1$

Area is: $TA_{right\ cylinder} = 2\pi r^2 + 2\pi rh$ units^2

Ratio $TA_{right\ cylinder} / TA_{right\ cylinder}$ is: $[32\pi + 64\pi] / [8\pi + 16\pi]$
$$= 96\pi / 24\pi = 4:1 = 2^2:1^2$$

Volume is: $V_{cylinder} = \pi r^2 h$ units^3

Ratio $V_{cylinder} / V_{cylinder}$ is: $[128\pi] / [16\pi] = 8:1 = 2^3:1^3$

(c)   Regular pyramids:

Perimeter ratio: $48/24 = 2:1$

Area is: $TA_{\text{regular pyramid}} = B + (1/2)p_{\text{base}}\,l$ units^2

Ratio $TA_{\text{regular pyramid}} / TA_{\text{regular pyramid}}$ is:

$[144 + (1/2)(48)(10)] / [36 + (1/2)(24)(5)] = 384 / 96 = 4:1 = 2^2:1^2$

Volume is: $V_{\text{pyramid}} = (1/3)Bh$ units^3

Ratio $V_{\text{pyramid}} / V_{\text{pyramid}}$ is:

$[(1/3)1{,}152] / [(1/3)144] = 348 / 48 = 8:1 = 2^3:1^3$

(d)   Right cones:

Circumference $(2\pi r)$ ratio: $6\pi/4\pi = 3:2 = 1.5:1$

Area is: $TA_{\text{right cone}} = \pi r^2 + \pi rl$ units^2

Ratio $TA_{\text{right cone}} / TA_{\text{right cone}}$ is: $[9\pi + 28.5\pi] / [4\pi + 12.6\pi]$

$= 37.5\pi / 16.6\pi \approx 2.25:1 = 1.5^2:1^2$

Volume is: $V_{\text{cone}} = (1/3)\pi r^2 h$ units^3

Ratio $V_{\text{cone}} / V_{\text{cone}}$ is:

$[(1/3)81\pi] / [(1/3)24\pi] = 27 / 8 = 3.375:1 = 1.5^3:1^3$

In each case (a) through (d), the ratio of corresponding areas is $x^2 : y^2$, and the ratio of corresponding volumes is $x^3 : y^3$.

• **Example:** If the heights of two rectangular solids are 18 and 30, and the surface area and volume of the smaller solid is 405 square units and 324 cubic units, respectively, what is the surface area and volume of the larger solid?

We can find the surface area of the larger solid using the *similar solids surface area theorem*: The ratio of the surface areas of two similar solids is equal to the *square of the ratio* of any pair of corresponding lengths.

(405 square units / x square units) $= (18$ units $/ 30$ units$)^2$

$405/x = (3/5)^2$

$405/x = 9/25$

$(405)(25) = 9x$

$10{,}125/9 = x = 1{,}125$ square units

Therefore, the surface area of the larger solid is 1,125 square units.

We can find the volume of the larger solid using the *similar solids volume theorem*: The ratio of the volumes of two similar solids is equal to the *cube of the ratio* of any pair of corresponding lengths.

(324 cubic units / y cubic units) = (18 units / 30 units)^3

324/y = (3/5)^3

324/y = 27/125

(324)(125) = 27y

40,500/27 = y = 1,500 cubic units

Therefore, the volume of the larger solid is 1,500 cubic units.

7.8 Cavalieri's Principle

• If a right rectangular prism is divided horizontally into thin rectangular slices, each rectangular slice would have the same area as the bases of the prism. If the slices are stacked in a slanted fashion, the volume is the same as it was when they were stacked straight.

• If the bases of two prisms lie in the same plane and all parallel planes in between the base planes cut cross sections with congruent areas, then the volumes of the two prisms are congruent.

Cavalieri's principle (postulate): For two geometric solids in a plane, if every plane parallel to the base plane that intersects one of the solids also intersects the other solid at the same distance from the base, and the resulting cross sections have the same area, then the two solids have the same volume.

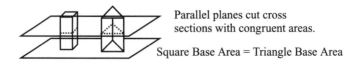
Parallel planes cut cross sections with congruent areas.

Square Base Area = Triangle Base Area

• Cavalieri's principle allows us to determine the volume of an oblique prism because if the base areas are equal and the heights are equal, then the volume is V = Bh regardless of overall shape. It follows that the volume formulas for pyramids, cylinders, and cones apply to both right and oblique solids.

• **Example:** If both cylinders have a diameter d of 8 feet and a height h of 12 feet, what are their volumes?

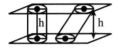

If diameter d = 8, then radius r = 4. The volume of a cylinder is given by:

$$V_{cylinder} = (\text{base area B})(h) = \pi r^2 h \text{ units}^3$$
$$= \pi(4 \text{ feet})^2(12 \text{ feet}) = 192\pi \text{ feet}^3 \approx 603 \text{ cubic feet}$$

Both cylinders have a volume of 192π cubic feet, because for a given base area and height the volume is the same regardless of shape.

7.9 Chapter 7 Summary and Highlights

• *Three-dimensional objects*, or *solids*, take up space in three dimensions. Measurements of three-dimensional objects include *volume* and *surface area*, as well as diagonal lengths. A *polyhedron* is a three-dimensional solid figure that is bounded by intersecting planes.

• *Prisms* are three-dimensional objects, or solids, having two congruent polygonal bases that lie in parallel planes. Examples of prisms include cubes, rectangular solids, triangular prisms, and pentagonal prisms. A *right prism* is a prism whose lateral edges are perpendicular to the two bases. The *surface area* TA of a right prism is equal to the area of the two congruent bases, 2B, plus the lateral area LA, which is the perimeter of a base p multiplied by the prism's height h:

$$TA_{right prism} = 2B + LA_{right prism} = 2B + ph \text{ units}^2.$$

The *volume* V of any prism is Bh, where B is the area of one of its bases and h is the length of the prism's altitude: $V_{prism} = Bh \text{ units}^3$.

• *Pyramids* are three-dimensional objects that have a square, rectangle, triangle, or other polygon base in a plane connected to a point outside the plane. In *regular pyramids*, the bases are *regular polygons* and the lateral edges are congruent. The *surface area* TA of a regular pyramid is equal to the area B of the base plus lateral area LA = (1/2)pl (having slant height l and perimeter p of the base):

$$TA_{regular pyramid} = B + LA_{regular pyramid} = B + (1/2)pl \text{ units}^2.$$

The *volume* V of any pyramid with base area B and altitude h is given by: $V_{pyramid} = (1/3)Bh \text{ units}^3$.

• **Cylinders,** or *circular solids,* are three-dimensional objects that have two identical circle bases in parallel planes connected by the set of all line segments between the two circle bases. If the segments joining the centers of the circles are perpendicular to the two planes of the circles, the cylinder is a *right circular cylinder.* The *surface area* TA of a right circular cylinder is equal to the area B of the circle bases plus the lateral area, $LA = Ch = 2\pi rh$, (where h is the altitude and $C = 2\pi r$ is the perimeter of a base):

$$TA_{\text{right cylinder}} = 2B + LA_{\text{right cylinder}} = 2B + Ch = 2\pi r^2 + 2\pi rh \text{ units}^2.$$

The *volume* V of any circular cylinder having two congruent circle bases with area $B = \pi r^2$ and altitude h is: $V_{\text{cylinder}} = Bh = \pi r^2 h \text{ units}^3$.

• **Cones** are three-dimensional objects that have a circle base connected to a vertex point outside the plane of the base. When the segment joining the center of the circle base and vertex point is perpendicular to the base, the cone is a *right circular cone.* The *surface area* TA of a right circular cone is equal to the lateral area, $LA = (1/2)2\pi rl = \pi rl$, (having slant height *l*), plus the circle *base area* B, which is the area of a circle $= \pi r^2$:

$$TA_{\text{right cone}} = B + LA_{\text{right cone}} = \pi r^2 + \pi rl \text{ units}^2.$$

The *volume* V of a circular cone having a circle base with area $B = \pi r^2$ and altitude h is given by: $V_{\text{cone}} = (1/3)Bh = (1/3)\pi r^2 h \text{ units}^3$.

• **Spheres** *or spherical solids* are three-dimensional objects consisting of points that are all the same distance from the center. The surface area A of a sphere with radius r is:

$$A_{\text{sphere}} = 4\pi r^2 \text{ units}^2.$$

The volume V of a sphere with radius r is: $V_{\text{sphere}} = (4/3)\pi r^3 \text{ units}^3$.

• **Similar solids,** like similar polygons, have the same shape but not necessarily the same size. If two solids are similar, the bases are similar and their corresponding lengths are in proportion. If the *ratio of lengths* is x:y, then the *ratio of areas* is $x^2:y^2$, and the *ratio of volumes* is $x^3:y^3$.

Chapter

8

Constructions and Loci

Plane geometry is the study of figures that can be constructed only with a straightedge and compass, and constructions are geometric drawings made using a straightedge and a compass.

8.1 Introduction

• Constructions are geometric drawings made using a straightedge and a compass. A straightedge is used to draw lines, and a compass is used to draw circles and arcs (which are sections of a circle). Arcs are used to locate points and points are used to determine lines.

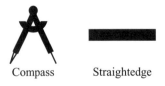

Compass   Straightedge

The length measure set by a **compass** is referred to as a **radius** (as in the radius of a circle), because compasses are used to draw circles and arcs of circles. The non-drawing point of a compass is used as the center of a circle or arc. A **straightedge** is similar to a ruler, however, it has no dimensional markings. The compass and straightedge are the tools of *plane geometry*.

• Working through a construction will encourage creative and visual thinking given that we have only two tools to use to identify such features as the perpendicular bisector of a line, the bisector of an angle, a perpendicular to a line, and a tangent to a circle. In addition, working through constructions teaches us many of the extraordinary properties of geometric figures, such as the following: Perpendicular bisectors of chords in a circle pass through the center of the circle; for any point on a circle, the tangent line and the radius line are perpendicular to each other at that point; and an angle inscribed in a semicircle forms a right angle.

The perpendicular bisectors of the *sides* of a triangle intersect at a point that is both equidistant from the three vertices of the triangle and at the center of a circle that circumscribes the triangle.

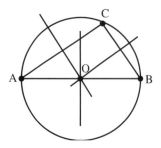

Additional properties revealed in constructions include: The bisectors of the *angles* of a triangle intersect at a point that is both equidistant from the three sides of the triangle and at the center of a circle that is inscribed in the triangle; and the area of a triangle is (1/2)(base)(altitude).

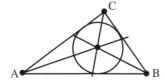
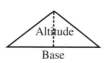

- The process of creating a construction involves:

1. A statement of what is to be constructed.
2. What is given.
3. What is to be constructed.
4. The procedure of the construction, which is a series of steps with figures that build the construction.
5. The concluding statement of what has been constructed and any relevant postulates or principles.

8.2 Constructions Involving Lines and Angles

- **Constructions in this section include**: 1. Congruent line segments; 2. Congruent angles; 3. Perpendicular bisector of a line segment; 4. Perpendicular to a line; 5. Perpendicular to a line from an external point; 6. Bisector of an angle; 7. Parallel lines; 8. Angle measuring 60°; 9. Dividing a segment into congruent parts; 10. Dividing a segment into proportional parts; and 11. Constructing proportional segments.

Congruent Segments and Angles:

1. **Construct a *line* segment *congruent* (equal in length) to a given line segment.**

Given: A line segment *AB*.

Construct: Line segment *CD* that is congruent to segment *AB*.

Procedure:

1. Begin with given line segment *AB*.
2. Draw a new line *l* longer than segment *AB* using a straightedge.
3. Select point C on line *l*.
4. Measure radius (length) of segment *AB* using a compass.
5. Use that compass measure, place the compass center point on point C, and make an arc on line *l* at the *AB* measured radius.
6. Label where the arc crosses line *l* as point D.

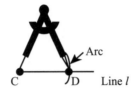

Therefore, new segment *CD* is congruent to original segment *AB*.

2. Construct an *angle congruent* (equal in measure) to a given angle.

Given: Angle A.

Construct: An angle congruent to ∠A.

Procedure:

1. Begin with given ∠A and draw a new ray that is similar to one side of angle A with its beginning point B.
2. Begin with ∠A and use a compass to draw an arc at any radius with its center point on point A.
3. Using the same compass radius measure, make a matching arc with its center point on B.

4. Using the compass, make a new arc that measures the distance between the two sides of angle A at the two points where the first arc crosses each side.
5. Using the same compass radius measure, make a matching new arc on ray B.

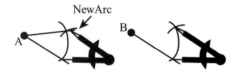

6. Draw a line from point B to where the two arcs intersect using the straightedge.

Angle B is congruent to angle A.

Note that if we use the arc crossings to create two triangles, they will be congruent with all corresponding sides equal (SSS). Because the sides all correspond, then $\angle A \cong \angle B$.

Bisectors and Perpendiculars:

3. **Construct the *perpendicular bisector* of a given *line* segment. (Bisect a line segment.)**

A perpendicular bisector is constructed by drawing two intersecting arcs from the endpoints. Then drawing a line through the arc-intersection points.

Given: A line segment *AB*.

Construct: The perpendicular bisector of *AB*.

Procedure:

1. Begin with given line segment *AB*.   A●————————●B
2. Use a compass with its center point first on point A, then on point B, and make two arcs across the line segment that intersect each other at two points above and below the line. (The two arcs should have the same radius measure and be slightly longer than half the length of the segment.) Label the two points where the arcs intersect as C and D.
3. Draw a line between points C and D using a straightedge.

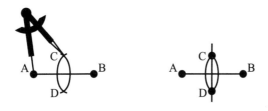

Therefore, line *CD* is the perpendicular bisector of line segment *AB*.

Note that points C and D are equidistant from A and B.

Relevant theorem: If a point is equidistant from the endpoints of a line segment, then the point lies on the perpendicular bisector.

Relevant postulate: Through any two points, there is one and only one line.

4. Construct a *perpendicular* to a given line through a given *point on the line*.

A perpendicular through a point on a line can be constructed by first drawing an arc from the point across both sides of the line. Then using those intersections, draw two more intersecting arcs through which the perpendicular is drawn to the point.

Given: Line *l* with point P on the line.

Construct: A perpendicular to line *l* through point P on line *l*.

Procedure:

1. Begin with line *l* containing point P.   —————•————— line *l*
2. Using a compass set with any radius and its center point on point P, draw an arc that intersects line *l* on each side of point P. Label the intersections point A and point B.

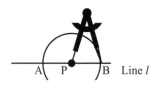

3. Using a compass with point A as the center and a radius greater than the distance between points A and P, draw an arc above point P. Then using the same compass radius and the center on point B, draw an arc above P that intersects the arc made from point A. Label where the arcs intercept as point C.
4. Draw a line between points C and P using a straightedge.

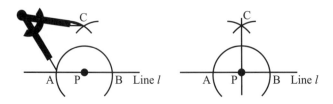

Therefore, line *CP* is the perpendicular line through line *l* at point P.

Note that points C and P are equidistant from A and B. Therefore, *CP* is a perpendicular bisector between A and B.

5. Construct a *perpendicular* to a given line from a given external point *not* on the line.

Given: Line *l* with point P *not* on the line.

Construct: A perpendicular to line *l* from external point P.

Procedure:

1. Begin with given line *l* and external point P.
2. Using a compass set with a radius greater than the distance between P and the line, and with its center point on point P, draw an arc that intersects line *l* on each side of point P. Label the intersections point A and point B.
3. Using a compass set with a radius greater than half the distance between points A and B, first use point A as the center and draw an arc below line *l* opposite P, then use point B as the center and draw an arc also below line *l* opposite P that crosses the first arc. Label where the arcs intercept as point C.
4. Draw a line between points P and C using a straightedge.

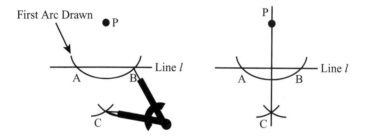

Therefore, line *PC* is the perpendicular to line *l* from external point P.

Note that points P and C are equidistant from A and B, therefore *PC* is a perpendicular bisector between points A and B.

6. Construct the *bisector* of a given *angle*.

Given: Angle A.

Construct: The bisector of angle A.

Procedure:

1. Begin with angle A.
2. Using a compass set with any radius and its center point on point A, draw an arc that intersects each side of angle A. Label the intersections point B and point C.

3. Using a compass with point B as the center and a radius greater than half the distance between points B and C, draw an arc beyond the BC arc. Then using the same compass radius and the center on point C, draw an arc that intersects the arc made from point B. Label where the arcs intercept as point D.
4. Draw a line between points A and D using a straightedge.

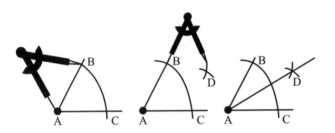

Therefore, line *AD* is the bisector through angle A.

Note that length *AB* = length *AC*, and the distance between points B and D is equal to the distance between points C and D.

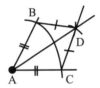

Note also that two congruent triangles, ΔABD and ΔACD, are formed (SSS postulate), with angle BAD corresponding to and being congruent to angle CAD. This confirms that line *AD* is the bisector of angle BAC.

Parallel Lines:

7. Construct a *line parallel* to a given line through a given external point *not* on the line.

Parallel lines can be constructed by creating congruent corresponding angles.

Given: Line *AB* and an external point P.

Construct: A line parallel to line *AB* through external point P.

Procedure:

1. Begin with given line *AB* and external point P.
2. Using a straightedge, draw a line through point P that crosses line *AB*. Label the point where the line crosses line *AB* as point C.

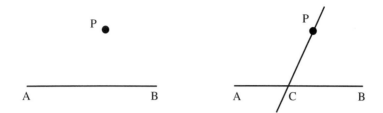

3. Using the compass, draw an arc with its center at point C across angle PCB. With the same compass radius, draw an identical arc with its center at P. Label the arc/line intersections as X, Y, and U.
4. Using the compass draw an arc with its center at point X where the lower arc crosses line *CP* and arc through point Y where the lower arc crosses line *AB*. With the same compass radius, draw an identical arc with its center at point U where the upper arc crosses line *CP*, and label the arc intersection as V.
5. Using a straightedge, draw a line labeled *DE* that spans through point P and point V where the two upper arcs intersect each other.

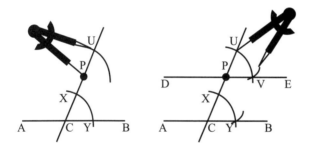

Therefore, line *AB* and line *DE* (through point P) are parallel.

Also, the indicated corresponding angles are congruent.

In this construction, we created congruent corresponding angles in order to construct a parallel line.

Relevant postulate: The *parallel transversal postulate*, which states that two parallel lines cut by a transversal have corresponding angles that are congruent.

8. Construct an *angle* of measure 60°

Given: Point A on line *AB*.

Construct: An angle of measure 60° using point A as the vertex.

Procedure:

1. Begin with given line *AB*.   A●————————●B
2. Using the compass with point A as the center, draw an arc through point B that extends up over line *AB* and back toward point A.
3. Using the same compass radius with point B as the center, draw an arc that intersects the first arc. Label the points where the arcs intersect as point C.
4. Using a straightedge, draw line segments between points B and C and between points A and C.

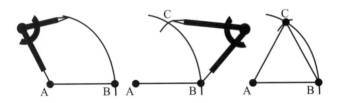

This creates an equilateral equiangular triangle with each angle measuring 60°. Therefore, creating angle A as a 60° angle.

This is true because the sum of the measures of the angles in all planar triangles is 180°, and 1/3 of 180° is 60°.

We have constructed the desired 60° angle by creating an equilateral triangle using arcs of equal radii.

Segments:

9. Construction: Divide a given *segment* into a given number of congruent parts.

Given: Line *AB*.

Construct: Divide line *AB* into three congruent parts by constructing two equidistant points (C and D) on segment *AB*.

Procedure:

1. Begin with given line segment *AB*.   A●————————●B
2. Extend a line *AE* from point A below segment *AB* using a straightedge.
3. Using a compass with point A as the center, draw three successive arcs across line *AE* that have equal radii, which are approximately 1/3 the length of segment *AB*.
4. Using the straightedge, draw a line through both point B and where the last arc crosses line *AE*.

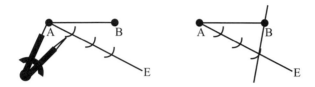

5. Where the first two arcs cross line *AE,* construct two lines parallel to the line that runs through both point B and the point of intersection of the last arc on line *AE.* These two parallel lines will intersect line *AB* at points C and D.

   These parallel lines are drawn by first making the larger arcs (depicted below) from the intersection points of the first arcs drawn on *AE.* Then use as centers where these large arcs cross *AE* and make intersecting arcs at the upper end of the large arcs, each with the same radius as where the farthest large arc intersects the first parallel line from point B. Draw lines through the first arcs on *AE* through the arc intersections. (See Construction 7 above, *Construct a line parallel to a given line through a given external point not on the line.*)

   Erase the arcs in order to visualize the congruent segments *AC,* *CD,* and *DB* that were created by the parallel lines.

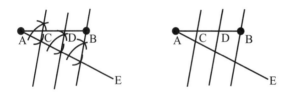

Therefore, segment *AB* is divided into three congruent segments of equal length: *AC, CD,* and *DB.*

10. Construction: Given two line segments, divide a third given line segment into *parts that are proportional* to the two segments.

To construct proportional line segments, we can use the principle that in a triangle, a line that intersects two of the sides and is parallel to the third side, will divide the two sides proportionally.

Given: Two line segments *AB* and *CD* and a third segment *XY.*

Construct: Divide the third given line segment *XY* into parts that are proportional to the two segments *AB* and *CD.*

Procedure:

1. Begin with given segments *AB* and *CD* and third segment *XY*. Draw a segment *XZ* that is at a convenient angle from *XY*.
2. Construct lengths *AB* and *CD* on *XZ* using a compass set with each of the radii for *AB* and *CD*. To do this, first set the center at point X, then use that arc as the center.

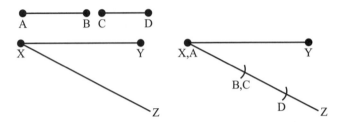

3. Draw a line from point D to point D to create triangle Δ*XYD*.
4. Construct a line parallel to *YD* through point B that intersects *XY* at point P.

The parallel line is drawn by first making large identical arcs from the points D and B. Then use as the centers where the large arcs cross *XZ*, and make intersecting arcs at the upper ends of the large arcs with the same radius as where the lower large arc intersects line *YD*. Draw a line from B through the arc intersection to P. (See Construction 7 above, *Construct a line parallel to a given line through a given external point not on the line*.)

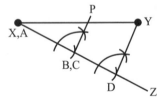

Therefore, line segments *XP* and *PY* are proportional to segments *AB* and *CD* using the principle that in a triangle, a line that intersects two of the sides and is parallel to the third side will divide the two sides proportionally.

11. Construction: Given three segments, construct a fourth segment so that the four *segments are in proportion*.

This is similar to the above construction and uses the same principle that in a triangle, a line that intersects two of the sides and is parallel to the third side will divide the two sides proportionally.

Given: Three line segments *AB*, *CD*, and *EF*.

Construct: A fourth segment so that the four segments are in proportion, and $AB/CD = EF/FP$.

Procedure:

1. Begin with given segments *AB*, *CD*, and *EF*, and draw a separate angle ∠YXZ that can contain the segments.
2. Construct lengths *AB* and *CD* on *XZ* using a compass set with each of the radii for *AB* and *CD*. To do this, first use X as the center point, then use that arc as the center.
3. Construct length *EF* on *XY* using a compass set with radius *EF*.

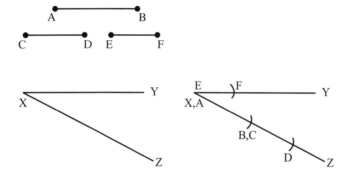

4. Draw a line from point F to point B to create triangle ΔXFB.
5. Construct a line parallel to *FB* through point D that intersects *XY* at point P.

The parallel line is drawn by first making identical large arcs from the points B and D. Then use as the centers where the large arcs cross *XZ*, and make identical intersecting arcs at the upper ends of the large arcs having the same radius as where the upper large arc intersects line *FB*. Draw a line from D through the arc intersection to P.

(See Construction 7 *Construct a line parallel to a given line through a given external point not on the line.*)

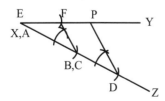

Therefore, the fourth segment *FP* is in proportion with segments *AB*, *CD*, and *EF* so that all four segments are in proportion, or *AB/CD = EF/FP*. This was achieved using the principle that in a triangle, ΔXPD, a line that intersects two of the sides and is parallel to the third side will divide the two sides proportionally.

8.3 Constructions Involving Triangles

• Constructions in this section include: 12. A triangle given its three sides; 13. A triangle given two sides and the included angle; 14. A triangle given two angles and the included side; 15. A right triangle given its hypotenuse and one leg; 16. An altitude of a triangle; 17. Congruent triangles; and 18. Similar triangles.

12. Construct a *triangle* given its three sides.

Given: Sides *AB*, *AC*, and *BC*.

Construct: Triangle ΔABC.

Procedure:

1. Begin with given sides *AB*, *AC*, and *BC*.
2. Start with line *AB*. With the compass, measure the radius of segment *AC*. Use that radius with A as the center point to draw an arc above *AB*.

3. With the compass, measure the radius of segment *BC*. Use that radius with B as the center point to draw an arc intersecting the first arc.
4. From where the two arcs intersect, draw lines to both A and B. Label the point where the arcs intersect as point C.

Therefore, we have constructed ΔABC using its three given sides *AB*, *AC*, and *BC*.

13. Construct a *triangle* given two sides and the included angle.

Given: Sides *AB* and *AC* and constructed angle A.

Construct: Triangle △ABC.

Procedure:

1. Begin with given sides *AB* and *AC* and constructed angle A on segment *AB*.

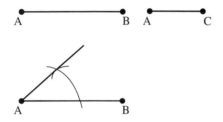

2. Using the compass with A as the center and a radius of *AC*, draw an arc across the upper side of angle A. Label this intersection point C.
3. Use the straightedge to draw a line between points B and C.

Therefore, we have constructed △ABC using two given sides *AB* and *AC* and included angle A.

14. Construct a *triangle* given two angles and the included side.

Given: Included side *AB* and angles A and B.

Construct: Triangle △ABC.

Procedure:

1. Begin with given side *AB* and construct angles A and B onto side *AB* with the angle vertices at the points A and B.
2. Extend the two sides using a straightedge to intersect each other at point C.

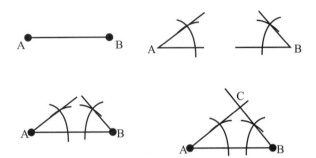

Therefore, we have constructed ΔABC using two given angles A and B and included side *AB*.

15. Construct a right *triangle* given its hypotenuse and one leg.

Given: The hypotenuse *AC* and leg *AB* of a triangle.

Construct: Right triangle ΔABC.

Procedure:

1. Begin with leg *AB* and hypotenuse *AC,* and using a compass set with a radius of *AB*, draw leg *AB* on a line *l*.
2. Construct a perpendicular of line *l* at point B. To do this, use a compass set with any radius and its center point on point B and draw equal-radius arcs that intersect line *l* on each side of point B.

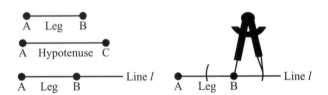

3. Using a compass with one arc-intersection point as the center and a sufficient radius, draw an arc above point B. Then using the same compass radius and the center on the other arc-intersection point, draw an arc above B that intersects the first arc.
4. Draw a line through the intersection point of the two arcs and point B using a straightedge.
5. Using a compass with a radius of *AC* (the hypotenuse) and point A as the center, draw an arc that just meets and intersects the perpendicular line (drawn from point B). Label the intersection C.
6. Draw hypotenuse *AC* using a straightedge.

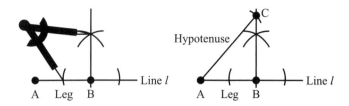

Therefore, we have constructed a right triangle given its hypotenuse and one leg.

16. Construct an *altitude* of a given *triangle*.

Note that an *altitude* of a triangle is a line that extends from one of the vertexes to the opposite side (or an extension of that side for obtuse triangles) that is perpendicular to that side. This construction is similar to Construction 5. *Construct a perpendicular to a given line from a given external point not on the line.*

Given: △ABC.

Construct: Altitude from vertex B.

Procedure:

1. Begin with given acute △ABC and obtuse △ABC.
2. Using a compass set with a radius greater than the distance between B and the opposite side (or an extension of the opposite side), and with its center point on B, draw an arc that intersects the opposite side at two points.

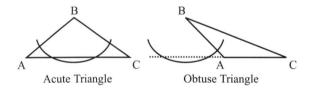

Acute Triangle   Obtuse Triangle

3. Set a compass with a radius greater than half the distance between the two points of intersection. First use the left intersection point as the center and draw an arc below the side opposite B (side *AC*), then use the right intersection point as the center and draw an arc also below the side opposite B that crosses the first arc.
4. Using a straightedge, draw a line from vertex B through where the arcs intercept below side *AC*. Label where the line crosses side *AC* as point D.

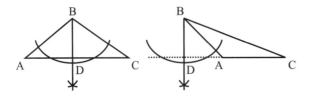

Line *BD* is perpendicular to side *AC* and extends from vertex B. Therefore, it is the altitude of ΔABC from vertex B.

Note that an altitude can be drawn from any vertex using this procedure.

17. Construct a *triangle congruent* to a given triangle.

Given: ΔABC and line *l*.

Construct: Congruent ΔDEF.

Procedure:

1. Begin with given ΔABC and line *l* labeled with point D.
2. Using a compass, measure side *AC* and use this radius with point D as the center to draw an arc through line *l*. Label the point where the arc crosses line *l* as F.

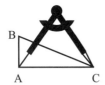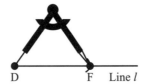

3. Using a compass, measure side *AB* and use this radius with point D as the center to draw an arc above point D.
4. Using the compass, measure side *BC* and use this radius with point F as the center to draw an arc that intersects the arc drawn in #3. Label the point of intersection of the two arcs as E.
5. Using a straightedge, draw lines *ED* and *EF*.

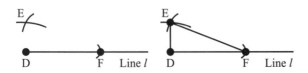

Therefore, we have constructed ΔDEF that is congruent to ΔABC with sides *AB* = *DE*, *AC* = *DF*, and *BC* = *EF*. These SSS congruent triangles can be written, ΔABC ≅ ΔDEF.

18. Construct a *triangle similar* to a given triangle given one side of the similar triangle.

Given: ΔABC and side *DE*.

Construct: Congruent triangle ΔDEF.

Procedure:

1. Beginning with given ΔABC and side *DE*, construct ∠D congruent to ∠A and construct ∠E congruent to ∠B. (See Construction 2.)

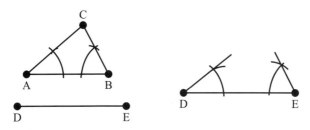

2. Using a straightedge, extend the two sides from ∠D and ∠E until they intersect each other at a point. Label that point as F.

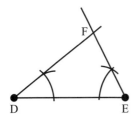

Therefore, we have constructed ΔDEF similar to ΔABC. These triangles have ∠A ≅ ∠D, ∠B ≅ ∠E, and ∠C ≅ ∠F.

Relevant postulate: AA similarity triangle postulate, which states that if two angles of one triangle are congruent to two angles of another triangle, then the triangles are *similar*.

8.4 Constructions Involving Circles and Polygons

•   Constructions in this section include: 19. Locate center of a circle; 20. A tangent to a circle; 21. A tangent to a circle from an external point; 22. Circumscribe a circle about a triangle; 23. Inscribe a circle in a triangle; 24. Inscribe a regular hexagon in a circle; 25. Inscribe an equilateral triangle in a circle; 26. Inscribe a square in a circle; and 27. Inscribe a regular octagon in a circle.

19. **Construction: Locate the** *center* **of a given** *circle*.

This construction uses the fact that a *perpendicular bisector* drawn toward the center of a circle from any *chord* in a circle will pass through the center of the circle. Because of this, we can draw two chords in a circle, and draw the perpendicular bisectors of those chords. Where they intersect will be the center of the circle. A chord is a line segment between any two points on a circle.

 The two line segments are chords of the circle.

Given: A circle.

Construct: Locate the center of a circle using the intersection of perpendicular bisectors of two chords.

Procedure:

1. Begin with the circle and draw two chords labeled *AB* and *CD*.
2. Draw a perpendicular bisector to chord *AB*: Using a compass with its center point first on point A, then on point B, make two arcs across the chord that intersect each other at two points above and below the chord. (The two arcs should have the same radius measure and be slightly longer than half the chord segment.)
3. Draw a line between the two points where the two arcs intersect using a straightedge.
4. Repeat numbers 2 and 3 for chord *CD*, drawing a perpendicular bisector to it: Using a compass with its center point first on point C, then on point D, make two arcs across the chord that intersect each other at two points above and below the chord. (The two arcs should have the same radius measure and be slightly longer than half the chord.)
5. Draw a line between the two points where the two arcs intersect using a straightedge.
6. Label the point where the two perpendicular bisectors intersect as P.

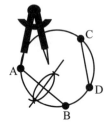 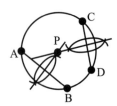

Therefore, the point of intersection of the two perpendicular bisectors of chords *AB* and *CD* is the center of the circle.

20. Construct the *tangent* to a given *circle* through a given point on the circle.

For a point on a circle, the tangent line and the radius line are perpendicular to each other at that point. Therefore, we can draw a line perpendicular to the radius line at the given point and it will be the tangent at that point.

Given: A circle C (having center C) with a given point P on the circle.

Construct: The tangent to the circle at point P.

Procedure:

1. Begin with circle C and the given point P on the circle, and draw the radius from center C to P and extend the radius line past P.
2. Using a compass set with any radius and its center point on point P, draw equal-radius arcs that intersect the radius line and its extension on each side of point P.
3. Using a compass with one arc-intersection point as the center and a sufficient radius, draw an arc above point P. Then, using the same compass radius and the center on the other arc-intersection point, draw an arc above P that intersects the first arc.
4. Draw a line between the intersection point of the two arcs and point P using a straightedge and label it line *t*.

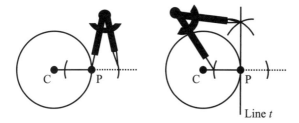

Line *t*

Therefore, we have constructed the tangent to a given *circle* through a given point on the circle using a perpendicular to a radius line.

Note that perpendicular lines form right angles and right angles form perpendicular lines. The following two lines are perpendicular to each other and the four angles formed at the intersection of the lines are right angles, which each measure 90°.

21. Construct a *tangent* to a given *circle* from a given external point outside the circle. Also, draw the second tangent from the point.

As we learned in the previous construction, for any point on a circle, the tangent line and the radius line are perpendicular to each other at that point. Because a tangent line will be perpendicular to a radius line at the point of tangency, we know that a tangent drawn from point P will touch the circle and be perpendicular to the radius line at that point. Other relevant information for this construction is that an angle inscribed in a semicircle forms a right angle. Therefore, we can use a circle to assist in forming a right angle between the tangent and radius.

Given: A circle C (having center C) with a point P *not* on the circle.

Construct: The tangents to the circle from given external point P.

Procedure:

1. Begin with circle C and point P *not* on the circle, and draw a line between center point C and external point P.
2. Draw a circle between points C and P, which will form two right triangles having vertices where the circles intersect. In order to draw a circle between points C and P, first bisect line *CP*. To draw the perpendicular bisector of segment *CP*, use a compass with its center point first on point C, then on point P, and make two arcs across segment *CP* that intersect each other at two points above and below the line. (The two arcs should have the same radius measure and be slightly longer than half segment *CP*.)
3. Draw a line between the two points where the two arcs intersect using a straightedge. Label the intersection of line *CP* and its perpendicular bisector as point O.
4. Using a compass with its center point as point O, draw a circle (dashed) with a radius of *OP* (and *OC*). (The diameter of the circle will be *CP*.)

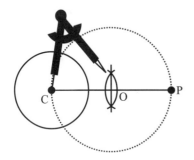

5. Label the two points A and B where circles C and O intersect.
6. Using a straightedge, draw lines connecting point P with the two points A and B. Doing this will form two right triangles whose hypotenuse is the diameter of circle O, or *CP*, and whose legs are *AP* and *BP* and also *AC* and *BC*. (An angle inscribed in a semicircle is a right angle, thereby giving perpendicular lines at its vertex where, in this example, a radius is perpendicular to a tangent.) Label the lines *l* and *t*.

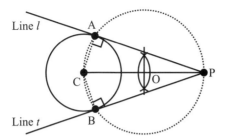

Therefore, we have constructed the two lines, *l* and *t*, tangent to the circle C from given external point P.

Related theorem: *Tangent segments theorem*, which states that tangent segments to a circle from an external point are congruent.

22. Construction: *Circumscribe* a *circle* about a given triangle.

The perpendicular bisectors of the sides of a triangle intersect at the center of a circle that circumscribes the triangle. The distance from the circle's center to each vertex is the radius of the circle.

Given: A triangle, ΔABC.

Construct: The circumscribed circle around ΔABC, which passes through the three vertex points of the triangle.

Procedure:

1. Begin with ΔABC and construct perpendicular bisectors to any two of its sides. First draw the perpendicular bisector of side *AB* using a compass with its center first on point A, then on point B, making two arcs of equal radii across side *AB* that intersect each other at two points above and below side *AB*. Then draw the bisector line between the two points where the two arcs intersect using a straightedge.
2. Draw the perpendicular bisector of side *AC* using a compass with its center first on point A, then on point C, making two arcs of equal radii across side *AC* that intersect each other at two points above and below side *AC*. Then draw the bisector line between the two points where the two arcs intersect using a straightedge.
3. Place point P where the bisectors intersect each other. Draw a circle using the compass with its center at point P. A radius of the circle is the measure between its center point P and any one of the three vertex points of the triangle.

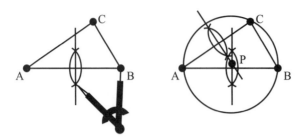

Therefore, we have circumscribed a circle about a given triangle.

Relevant theorem: The perpendicular bisectors of the sides of a triangle intersect at a point that is equidistant from the three vertices of the triangle.

23. Construction: *Inscribe* a *circle* in a given triangle.

The bisectors of the angles of a triangle intersect at the center of a circle that is inscribed in the triangle. The inscribed circle has a radius equal to the length of a line drawn perpendicular to any side from its center.

Given: A triangle, ΔABC.

Construct: The inscribed circle inside ΔABC, which touches each of the three sides of the triangle.

Procedure:

1. Begin with ΔABC and construct bisectors to any two of its angles. Using a compass set with any radius and its center point on point A, draw an arc that intersects each side of ∠A.
2. Using a compass with the center first on the arc-intersection of one side of A then on the arc-intersection of the other side of A, and a sufficient radius, draw two arcs that intersect each other. Draw a line between point A and the intersection of the two arcs using a straight-edge. This is the bisector of ∠A.

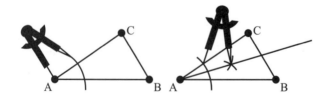

3. Repeat numbers 1 and 2 for ∠B. This will result in the bisector for ∠B. Label the intersection point of the two bisectors as point P.

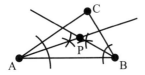

4. Draw a perpendicular to side *AB* from point P (the intersection of the two bisectors). To do this, first use a compass set with a radius greater than the distance between point P and side *AB* and with its center point P, draw an arc that intersects side *AB* at two points. Next, use the left intersection point as the center and draw an arc below side *AB*, then use the right intersection point as the center and draw an arc also below the side *AB* that crosses the first arc. Finally, using a straightedge draw the perpendicular line from point P across side *AB* to the point where the arcs intersect. Label the point where the perpendicular crosses side *AB* as point S.
5. Using a compass with its center on point P (where the angle bisectors intersect), draw a circle with a radius that is the perpendicular drawn from point P to point S (on side *AB*).

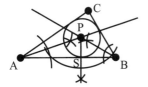

Therefore, we have inscribed a circle inside a triangle.

Relevant theorem: The bisectors of the angles of a triangle intersect at a point that is equidistant from the three sides of the triangle.

24. Construction: *Inscribe* a regular hexagon in a given *circle*.

A regular hexagon is a six-sided polygon in which all sides have the same length. When a regular hexagon is inscribed in a circle, each of its six vertices will touch the inside of the circle. From the center of the circle to each vertex is the radius of the circle. Because a hexagon divides a circle's 360° by 6, the result is that each triangular section between two radii has a *central angle* of 60°. In addition, because the angles in all planar triangles sum to 180°, these triangular regions in the hexagon are equilateral triangles, each having three 60° angles. In this special hexagon case, the radius of the circle is equal to the length of each side of the inscribed hexagon, which makes this construction easy.

Given: A circle C.

Construct: Inscribe a regular hexagon inside circle C (so that each vertex touches the inside of the circle).

Procedure:

1. Begin with circle C. If you don't know the center point of the circle use Construction 19, which locates the center of a given circle. (This construction uses the principle that a perpendicular bisector drawn toward the center of a circle from any chord in a circle, will cross the center of the circle. Because of this, we can locate the center of any circle by drawing two chords in the circle, then draw the perpendicular bisectors of those chords. Where they intersect will be the center of the circle. A *chord* is a line segment between any two points on a circle.)
2. Using a compass, measure the radius of the circle. Select a point (1) on the circle and make successive arcs around the circle using the same radius measure. The intersection of each arc becomes the new center point for drawing the next arc. Label the arc-intersections 1 through 6.
3. Using the straightedge, draw a line segment between each of the arc-circle intersection points (1, 2, 3, 4, 5, and 6). Drawing these segments will form the six sides of a regular hexagon.

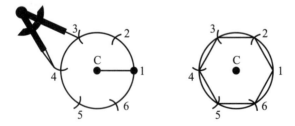

Therefore, we have inscribed a regular hexagon into a circle.

25. Construction: *Inscribe* an equilateral triangle in a given *circle*.

An equilateral triangle inscribed in a circle has its three vertices on the circle at an equal distance from each other. We can use the preceding Construction, 24, which inscribes a regular hexagon in a given circle, and instead of drawing line segments to each arc we join every other arc.

Given: A circle C.

Construct: Inscribe an equilateral triangle inside circle C (so that each vertex is on the circle).

Procedure:

1. Begin with circle C. (If you don't know the center point of the circle, use Construction 19, which locates the center of a given circle.)
2. Using a compass, measure the radius of the circle. Then select a point (1) on the circle and make successive arcs around the circle using the same radius measure. The intersection of each arc becomes the new center point for drawing the next arc. Label the arc-intersections 1 through 6.
3. Using the straightedge, draw a line segment between every other arc-circle intersection point (1 and 3, 3 and 5, 5 and 1) (or 2 and 4, 4 and 6, and 6 and 2). Drawing these segments will form the three sides of an equilateral triangle.

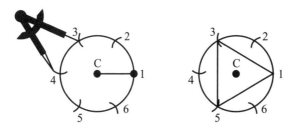

Therefore, we have inscribed an equilateral triangle in a circle using the first two steps for the construction of inscribing a regular hexagon in a circle (in which the radius of the circle was equal to the side of the hexagon).

26. Construction: *Inscribe* **a square in a given** *circle*.

We can inscribe a square in a circle by constructing two perpendicular diameter line segments and connecting their end points.

Given: A circle C.

Construct: Inscribe a square inside circle C (so that each vertex is on the circle).

Procedure:

1. Begin with circle C. If you don't know the center point of the circle use Construction 19, which locates the center of a given circle. This construction uses the principle that a perpendicular bisector drawn toward the center of a circle from any chord in a circle will cross the center of the circle. Because of this, we can locate the center

of any circle by drawing two chords in the circle, then drawing the perpendicular bisectors of those chords, and where they intersect will be the center of the circle. A *chord* is a line segment between any two points on a circle.

2. Draw a diameter through the center of the circle.

3. Construct a second diameter that is perpendicular to the first. To construct a perpendicular (see Construction 4, *Construct a perpendicular to a given line through a given point on the line*), begin with the first diameter line containing center point C, and using a compass set with any radius and its center point on point C, draw an arc that intersects the diameter line on each side of point C.

4. Using a compass with its center where the arc crosses one side of the diameter and a sufficient radius, draw an arc above point C. Then, using the same compass radius with the center on where the arc crosses the other side of the diameter, draw an arc above C that intersects the first arc.

 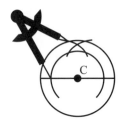

5. Using a straightedge, draw a diameter line across the circle that passes through the center point C and the point where the two arcs drawn in #4 intersect.

6. Join the endpoints of the two diameter lines with segments to form a square.

Therefore, we have inscribed a square inside a circle by connecting two perpendicular diameter lines.

27. Construction: *Inscribe* a regular octagon in a given *circle*.

A regular octagon can be inscribed in a circle by beginning with the construction of perpendicular diameters (as described in the preceding construction) and then bisecting the 90° angles formed by the perpendicular diameter lines.

Given: A circle C.

Construct: Inscribe a regular octagon inside circle C so that each vertex is on the circle.

Procedure:

1. Begin with circle C. If you don't know the center point of the circle use Construction 19, which locates the center of a given circle. Then, draw a diameter through the center of the circle.
2. Construct a second diameter that is perpendicular to the first. To construct a perpendicular (see Construction 4, *Construct a perpendicular to a given line through a given point on the line*), begin with the first diameter line containing center point C and using a compass set with any radius and its center on point C, draw an arc that intersects the diameter line on each side of point C.
3. Using a compass with its center where the arc crosses one side of the diameter and a sufficient radius, draw an arc above point C. Then, using the same compass radius with the center on where the arc crosses the other side of the diameter, draw an arc above C that intersects the first arc.
4. Using a straightedge, draw a diameter line across the circle that extends through the center point C and the point where the two arcs drawn in #3 intersect.

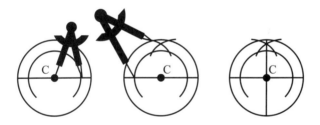

5. (Erase arcs.) Bisect the 90° angles formed by the perpendicular diameter line segments. Begin with the upper right angle, and using a compass set with any radius and its center on point C, draw an arc that intersects each side of the upper right angle.
6. Using a compass with its center first on one side of the angle where the arc from #5 intersects and then (with the same radius) and its center on the other side of the angle where the arc from #5 intersects, draw two arcs that intersect each other.
7. Draw a line from the point where the two arcs intersect through the center C of the circle and extend it to the far side of the circle. This line bisects both the upper right and lower left angles.

8. Repeat #5, #6, and #7 and bisect the upper left angle and extend the line to the lower right quadrant. This line bisects both the upper left and lower right angles. The result is eight congruent central angles and arcs.

9. Connect each point where the sides of the angles intersect the circle with line segments (or chords).

Therefore, we have inscribed a regular octagon inside a circle by constructing perpendicular diameters and bisecting the resulting four angles.

8.5 Construction Involving Area

• The construction in this section is a scalene and an isosceles triangle having the same base and area.

28. Construction: Given a scalene triangle, construct an isosceles triangle having the same base and equal in area.

A scalene triangle has its three sides of unequal length, and an isosceles triangle has two sides of equal length. Because the area of a triangle is (1/2)(base)(altitude), then if the triangles both have the same base, we can construct an isosceles triangle so that its altitude is the same length as the altitude of the scalene triangle.

Given: Scalene triangle ΔABC.

Construct: Isosceles triangle ΔABD having the same base and equal area.

Procedure:

1. Begin with scalene triangle ΔABC and construct a line through vertex C that is parallel to base *AB*. To construct a parallel line we create congruent corresponding angles. First, begin with side *AB* of given scalene triangle ΔABC and vertex point C. To find the angle that the parallel line makes with side *AC*, we can either extend side *AC* beyond vertex point C or we can use the principle which states that if two parallel lines are cut by a transversal, then alternate interior angles are congruent.

    Using the compass draw an arc with its center at point A across angle CAB. With the same compass radius, draw an identical (but upside-down-backwards) arc with its center at C.

2. Using a compass, draw an arc with its center at the point where the lower arc crosses line *AC* and arc through the point where the lower arc crosses line *AB*. With the same compass radius, draw an identical (upside-down-backwards) arc with its center at the point where the upper arc crosses line *AC*.

3. Then, using a straightedge, draw a line that spans through point C and the point where the two upper arcs intersect each other. This line through vertex C will be parallel to side *AB*. (Note that we can also construct the parallel line by using the extension of side *AC*. See Construction 7.)

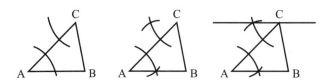

4. (Erase arcs.) Construct the perpendicular bisector of side *AB*. To do this, use a compass with its center first on point A, then on point B, making two arcs of equal radii across side *AB* that intersect each other at two points above and below the line. Then draw the bisector line between the two points where the two arcs intersect and up to the parallel line that was drawn through vertex C. Label the intersection of the perpendicular bisector and the parallel through C as point D.

5. Using a straightedge, draw a line from point A to point D and also from point B to point D. Because *AD* and *BD* are drawn to the same point on the perpendicular bisector between them, they are equal in length, and the triangle ΔABD is an isosceles triangle.

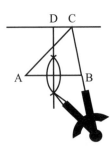
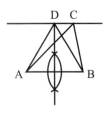

Because area = (1/2)(base)(altitude) and both triangles, scalene △ABC and isosceles △ABD, have the same base and the same altitude length, they have the same area. (They have the same altitude length because both altitudes fit exactly between parallel lines *DC* and *AB*.)

Relevant theorem: *Common base triangle area theorem*, which states that triangles have equal areas if they share a common base and their third vertices lie on a line that is parallel to their base.

8.6 Locus of Points

• The *locus* is the set of points that satisfies a certain condition or conditions. For example, a perpendicular bisector represents the locus of points equidistant between two points on the line connecting them. In this case, the locus of points of the perpendicular bisector satisfies the conditions that require all points to be equidistant between two given points.

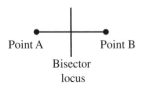

• Similarly, the locus of points that satisfies the condition that it is equidistant between two parallel lines is a third parallel line that is exactly between the first two parallel lines.

Line 1 ——————— } distance
Locus ——————— } distance
Line 2 ———————

• A locus of points can be in a plane or in three-dimensional space. For example, in a plane, the locus of points equidistant between two concentric circles is a circle concentric with the first two circles and halfway between them. In three-dimensional space, the locus of points equidistant between two concentric spheres is a sphere concentric with the first two spheres and halfway between them.

2D

3D

• *Loci* is the plural of locus. There may be more than one condition that must be satisfied and may involve more than one locus of points, such as the intersection of the locus of points forming a circle with the locus of points forming a plane. For example, the locus of points satisfying the condition that it is 2 miles from point P in a plane (forming a circle) and also the condition that it is 2 miles in a given direction from plane W. These two locus of points may or may not intersect each other at all, or the circle may be tangent to the plane or coplanar with the plane depending on the location of point P and plane W. Therefore, the locus of points that satisfies both conditions may be zero points, two points, one point, or a circle of points.

• Additional examples of locus include the following:

The locus of points satisfying the condition that it is equidistant from the sides of a given angle, is the bisector of that angle.

Locus is bisector of angle.

The locus of points satisfying the condition that it is a given distance from a point, is a circle with the point as its center and the distance as its radius.

Locus is circle.

The locus of points satisfying the condition that it is a given distance from a line in a plane, is two lines parallel to the given line at the given distance.

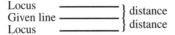

The locus of points satisfying the condition that it is a given distance from a line in 3-dimensional space, is a cylindrical surface parallel to and surrounding the given line at the given distance.

Locus is cylindrical surface.

• Examples of what a locus may be used to describe include: the path of an orbiting satellite, a boat traveling through the center of a canal, and a radar system depicting the locations of airplanes near an airport.

• Locus is also used in analytic, or coordinate, geometry (see Chapter 9) in which the locus is represented by a graph of an equation that is given by the points or coordinates that satisfy the equation. The locus is the set of points (or graph) of an equation that is given by the coordinates (x, y) or $((x, y, z)$ for three dimensions) that satisfy the equation. In other words, the locus, or graph, of an equation containing two variables, such as x and y, is the curve or line that contains all of the points, and no others, whose coordinates satisfy the equation. From Chapter 9:

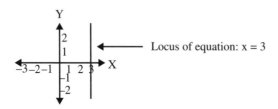

8.7 Chapter 8 Summary and Highlights

• *Plane geometry* is the study of figures that can be constructed only using a straightedge and compass. **Constructions** are geometric drawings made using a straightedge and a compass. A **straightedge** is used to draw lines and a **compass** is used to draw circles and arcs (which are sections of a circle). In a construction, arcs are used to locate points, and points are used to determine lines. The length measure set by a *compass* is referred to as a *radius* (as in the radius of a circle) because compasses are used to draw circles and arcs of circles. A *straightedge* is similar to a ruler; however, it has no dimensional markings. The compass and straightedge are the tools of *plane geometry.*

• The process of creating a construction involves: (1) A statement of what is to be constructed; (2) What is given; (3) What is to be constructed; (4) The procedure of the construction, which is a series of steps with figures that build the construction; and (5) The concluding statement of what has been constructed and relevant postulates or principles.

• This chapter explains the following constructions:

Constructions involving lines and angles: 1. Congruent line segments; 2. Congruent angles; 3. Perpendicular bisector of a line segment; 4. Perpendicular to a line; 5. Perpendicular to a line from an external point; 6. Bisector of an angle; 7. Parallel lines; 8. Angle measuring 60°; 9. Divide segment into congruent parts; 10. Divide segment into proportional parts; and 11. Construct proportional segments.

Constructions involving triangles: 12. A triangle given its three sides; 13. A triangle given two sides and the included angle; 14. A triangle given two angles and the included side; 15. A right triangle given its hypotenuse and one leg; 16. An altitude of a triangle; 17. Congruent triangles; and 18. Similar triangles.

Constructions involving circles and polygons: 19. Locate center of a circle; 20. A tangent to a circle; 21. A tangent to a circle from an external point; 22. Circumscribe a circle about a triangle; 23. Inscribe a circle in a triangle; 24. Inscribe a regular hexagon in a circle; 25. Inscribe an equilateral triangle in a circle; 26. Inscribe a square in a circle; and 27. Inscribe a regular octagon in a circle.

Constructions involving area: 28. Construct an isosceles triangle having the same base and equal in area as a scalene triangle.

• The *locus* is the set of points that satisfies a certain condition or conditions. A locus of points can be in a plane or in three-dimensional space. For example in a plane, the locus of points equidistant between two concentric circles is a circle, and in three-dimensional space, the locus of points equidistant between two concentric spheres is a sphere.

Chapter

9

Coordinate or
Analytic Geometry

Euclidean geometry combines related elements using the methods of logic and reasoning and the tools of axioms, postulates, definitions, theorems, and constructions in order to prove, describe, calculate, generate, or use information pertaining to geometric objects. *Coordinate geometry,* also called *analytic geometry*, was created by René Descartes in the 17th century as a new geometry that combined Euclidean geometry with algebra. **Coordinate,** or **analytic geometry**, is the study of geometry using the analytical methods of algebra. This approach involves placing a geometric figure into a coordinate system illustrating the proof, and obtaining information about the figure using algebraic equations.

Graphing on a coordinate system is often used to visualize quantitative data in a manner that will provide insight into trends, patterns, or relationships. Graphing an equation on a coordinate system provides a depiction of the slope in the case of a linear equation, or the shape of a curve in the case of a nonlinear equation. Geometrical figures defined by equations can also be depicted on a coordinate system. Coordinate, or analytic, geometry is used in everything from air traffic control, geometric proofs, and calculating lengths, distances, and slopes, to plotting the speed, or velocity, of a bicyclist.

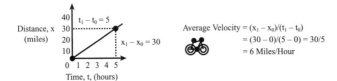

9.1 Rectangular Coordinate Systems: Definitions

• *Real numbers* can be identified with points on a number line and depicted on the ***real number line***. A number line is a ***one-dimensional coordinate system***. All real numbers, except zero, are either positive or negative. All real numbers correspond to points on the real number line, and all points on the number line correspond to real numbers. The real number line reaches from negative infinity ($-\infty$) to positive infinity ($+\infty$).

$$\xleftarrow{\hspace{1cm}}\quad -4 \qquad -3 \qquad -2 \qquad -1 \;\; -.5 \;\; 0 \qquad 1 \;\; \sqrt{2} \;\; 2 \;\; 5/2 \;\; 3 \;\; \pi \qquad 4 \xrightarrow{\hspace{1cm}}$$

The real number line includes the real numbers -0.5, -2, $5/2$, and $\pi = 3.14159...$ Remember, the *real number system* is comprised of irrational numbers and rational numbers, including natural numbers, whole numbers, integers, fractions, and decimals.

• If a number is represented by a point on the number line, the number is called the ***coordinate*** of that point. A number can be represented in one, two, or three dimensions. Pairs of real numbers that define a point on a plane can be depicted by identifying them with the two axes of a *two-dimensional coordinate system*. Points in three-dimensional space can be depicted by identifying them with the three axes of a *three-dimensional coordinate system*.

• A ***point*** on the number line is represented in one dimension. On a number line, the numbers to the right are greater than the numbers to the left. Therefore, a number to the left of another number is less than a number to the right. To define the position of a point on a number line, the number the point corresponds to is identified.

$$\xleftarrow{\hspace{1cm}} \bullet \quad -4 \;\; -3 \;\; -2 \;\; -1 \;\; 0 \;\; 1 \;\; 2 \;\; 3 \;\; 4 \xrightarrow{\hspace{1cm}}$$

The point is at the -3 coordinate position on the number line.

• To ***add or subtract negative and positive numbers***, we can think of moving along the number line (depicted below). Begin with the first number and move to the *right for positive numbers or addition*, or move to the *left for negative numbers or subtraction*.

If you are at –3 and you want to add 3, you move to the right
   3 places and end up at 0: –3 + 3 = 0.

If you are at 4 and you want to add 1, you move to the right
   1 place and end up at 5: 4 + 1 = 5.

If you are at 5 and you want to subtract 7, you move to the left
   7 places and end up at –2: 5 – 7 = –2.

If you are at –3 and you want to subtract –4, you move to the *right*
   4 places and end up at 1: –3 – (–4) = –3 + 4 = 1.

Remember, when you subtract a negative number the two minus signs cancel each other and you end up with a plus sign.

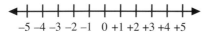

The distance between zero and a number on the number line is called the ***absolute value*** or the ***magnitude*** of the number. The absolute value of a number is always positive or zero, never negative. The symbol for absolute value of a number represented by n is $|n|$. For example, positive 4 and negative 4 have the same distance from zero, or an absolute value of: $|4| = 4$ and $|-4| = 4$.

• The ***distance between two points on the number line*** *is the absolute value of the difference in their coordinates*. The distance between two points is also the *length of the line segment between them*.

What is the distance between –2 and 2? Subtract the numbers.
   $|-2 - 2| = |-4| = 4$, or equivalently $|2 - (-2)| = |2 + 2| = 4$

The length of the segment between –2 and 2 is 4.

What is the distance between –1 and 2? Subtract the numbers.
   $|-1 - 2| = |-3| = 3$, or equivalently $|2 - (-1)| = |2 + 1| = 3$

The length of the segment between –1 and 2 is 3.

What is the distance between –5 and –1? Subtract the numbers.
   $|-5 - (-1)| = |-5 + 1| = |-4| = 4$, or $|-1 - (-5)| = |-1 + 5| = 4$

The length of the segment between –5 and –1 is 4.

• Two-dimensional, planar *rectangular coordinate systems* (also called *Cartesian coordinate systems*) consist of two *axes*, generally denoted X and Y. X represents the horizontal axis and Y the vertical axis. X is sometimes called the *abscissa* and Y the *ordinate*. The axes are at *right angles, or perpendicular,* to each other. The axes intersect at zero on the X-axis and zero on the Y-axis. The center of the coordinate system where the axes intersect is called the *origin*. (See *Master Math: Algebra*.)

• A *point* on a planar rectangular coordinate system is represented in two dimensions, x and y. To define the position of a point on a two-axis planar coordinate system, the numbers on each axis that the point corresponds to are the coordinates of the point with respect to that axis.

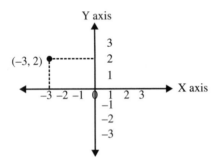

The point corresponds to –3 on the X-axis and +2 on the Y-axis. The point's coordinates are (–3, 2).

• Each point on an X-Y coordinate system corresponds to a unique *ordered pair* of real numbers (x, y), where x and y are the *coordinates of the point*. The convention for writing numbers that a point corresponds to (the *ordered pair*) on each axis is:

(X-axis-number, Y-axis-number)

In the preceding coordinate system, (–3, 2) is the coordinate of the point depicted.

The x-coordinate is the first number in the ordered pair, and corresponds to the location of the point along the X-axis.

The x-coordinate is positive if it is on the right side of the Y-axis and negative if it is on the left side of the Y-axis.

The y-coordinate is the second number in the ordered pair, and corresponds to the location of the point along the Y-axis.

The y-coordinate is positive if it is above the X-axis and negative if it is below the X-axis.

• A point represented by (x, y) on an X-Y coordinate system can be identified as being located in a particular quadrant. The axes of a planar Cartesian coordinate system divide the plane into *quadrants* I, II, III, and

IV, which are typically labeled using Roman numerals counterclockwise
beginning with the top right quadrant I.

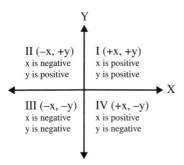

- **Example**: What are the coordinates of the vertices of the pentagon in
the graph below?

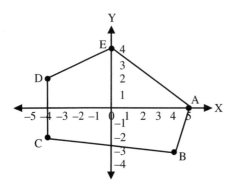

Each point on an X-Y coordinate system corresponds to a unique ordered
pair of real numbers, (x, y), where x and y are the coordinates of the point,
(X-axis-number, Y-axis-number). The coordinates of the points in the
graph are:

    A(5, 0), B(4, –3), C(–4, –2), D(–4, 2), and E(0, 4).

- A *three-dimensional rectangular, or Cartesian, coordinate system*
consists of three axes, generally denoted x, y, and z, which are all three at
right angles to each other. A *point on a spatial three-dimensional rectangular coordinate system* (see figure below) with three intersecting axes
that are all perpendicular to each other is represented in three dimensions.
X represents the horizontal axis, Y represents the vertical axis, and Z
represents the axis that comes out of the page and is perpendicular to the
page with the positive side of the axis in front of the page and the negative
side of the axis behind the page. The point is represented in three-

dimensional space. The axes intersect at zero on the X-axis, Y-axis, and Z-axis. To define the position of a point on a three-axis (three-dimensional) coordinate system, the number on each axis that the point corresponds to is identified.

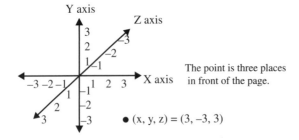

The point corresponds to x = +3 on the X-axis, y = –3 on the Y-axis, and z = +3 on the Z-axis, and can be identified by coordinates (x, y, z) = (3, –3, 3). (The position of the point with respect to the Z-axis cannot be accurately visualized in two dimensions.)

9.2 Distance Between Points

• What is the shortest *distance between two points*? A straight line. The distance between points (x_1, y_1) and (x_2, y_2) on a coordinate system is given by: $d^2 = (x_2 - x_1)^2 + (y_2 - y_1)^2$

This formula can be obtained from the Pythagorean Theorem for a right triangle. First, consider the distance between two points that are along a horizontal or vertical line, parallel to either axis. In the figure below, what is the distance between points A and B or points B and C?

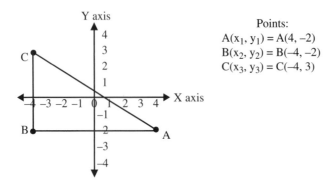

Points:
$A(x_1, y_1) = A(4, -2)$
$B(x_2, y_2) = B(-4, -2)$
$C(x_3, y_3) = C(-4, 3)$

Determining the distance between points in a horizontal or vertical line involves simply moving along the corresponding axis, which is a number line, and taking the difference between the numbers. Remember from the last section: The distance between points on a number line is the absolute value of their difference, and also the length of the line segment between them.

For points A and B, the distance along the X-axis is:

Segment $AB = |4 - (-4)| = |4 + 4| = 8$, or equivalently,
$|-4 - 4| = |-8| = 8$

For points B and C, the distance along the Y-axis is:

Segment $BC = |3 - (-2)| = |3 + 2| = 5$, or equivalently,
$|-2 - 3| = |-5| = 5$

Therefore, the distance $AB = 8$ and the distance $BC = 5$.

Now suppose we want to know the distance between points A and C, (or the length of segment AC). To determine AC, we can extend line segments parallel to each axis to create a right triangle, (which is already on the previous graph as segments AB and BC). To find the distance AC, we use the *Pythagorean Theorem*, $\text{leg}^2 + \text{leg}^2 = \text{hypotenuse}^2$, for $\triangle ABC$:

$AB^2 + BC^2 = AC^2 = 8^2 + 5^2 = 89$

$AC = \sqrt{89} \approx 9.4$.

• A faster way to determine the distance between two points in a coordinate system is to use the *distance formula*: $d^2 = (x_2 - x_1)^2 + (y_2 - y_1)^2$.
We can derive the distance formula using the Pythagorean Theorem. In the figure, point A = $A(x_1, y_1)$, point B = $B(x_2, y_2)$, and point C = $C(x_3, y_3)$, or as depicted in the graph: $A(x_1, y_1) = A(4, -2)$, $B(x_2, y_2) = B(-4, -2)$, and $C(x_3, y_3) = C(-4, 3)$.

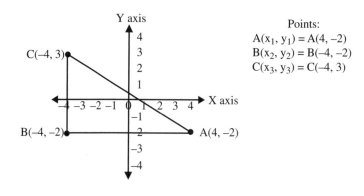

Points:
$A(x_1, y_1) = A(4, -2)$
$B(x_2, y_2) = B(-4, -2)$
$C(x_3, y_3) = C(-4, 3)$

The distance between any two points is the length of the line segment between them, so we can write:

$AB = (x_2, y_2) - (x_1, y_1)$, which represents AB in terms of x_1 and x_2, and y_1 and y_2.

$x_2 - x_1 = |-4 - 4| = |-8| = 8$, and

$y_2 - y_1 = |-2 - (-2)| = |-2 + 2| = 0$

Because $y_2 - y_1 = 0$, then distance AB is just $x_2 - x_1$.

$BC = (x_3, y_3) - (x_2, y_2)$, which represents BC in terms of x_3 and x_2, and y_3 and y_2.

$x_3 - x_2 = |-4 - (-4)| = |-4 + 4| = 0$, and

$y_3 - y_2 = |3 - (-2)| = |3 + 2| = 5$

Because $x_3 - x_2 = 0$, then distance BC is just $y_3 - y_2$.

We find the distance AC using *Pythagorean Theorem* as it corresponds to the axes:

$AC^2 = AB^2 + BC^2 \rightarrow AC^2 = (x_2 - x_1)^2 + (y_3 - y_2)^2$

$AC^2 = (|-4 - 4|)^2 + (|3 - (-2)|)^2 = 8^2 + 5^2 = 89$

$AC = \sqrt{89} \approx 9.4$,

which we previously determined using the Pythagorean Theorem directly.

We have derived the ***distance between two points formula*** for two points A and C in a coordinate system, where the distance d, or AC, is given by this formula for the *distance between two points*:

$d^2 = (x_2 - x_1)^2 + (y_2 - y_1)^2$

• ***Distance between two points formula theorem***: If the coordinates of two points are (x_1, y_1) and (x_2, y_2), then the distance d between the two points is given by the ***distance formula***:

$d^2 = (x_2 - x_1)^2 + (y_2 - y_1)^2$

• In summary, to determine the *distance between points* in a coordinate system, a right triangle can be created by extending lines parallel to the axes and using the Pythagorean Theorem to calculate the distance, which is the length of the hypotenuse. Alternatively, (and more quickly), the points can be defined by the X and Y axes of the coordinate system, and the distance d between points (x_1, y_1) and (x_2, y_2) can be calculated using the distance formula: $d^2 = (x_2 - x_1)^2 + (y_2 - y_1)^2$

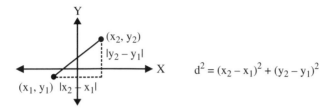

$$d^2 = (x_2 - x_1)^2 + (y_2 - y_1)^2$$

• *Side note*: The distance d between two points in three-dimensional space in a three-dimensional coordinate system is represented using:

$$d^2 = (x_2 - x_1)^2 + (y_2 - y_1)^2 + (z_2 - z_1)^2$$

• **Example**: Prove that the diagonals of a rectangle are congruent using a rectangle with vertices: A(4, 3), B(4, 0), C(0, 0), and D(0, 3).

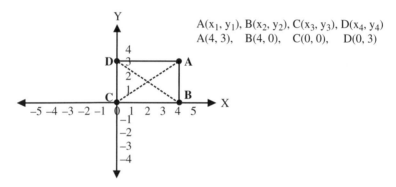

$A(x_1, y_1)$, $B(x_2, y_2)$, $C(x_3, y_3)$, $D(x_4, y_4)$
A(4, 3), B(4, 0), C(0, 0), D(0, 3)

In rectangle ABCD, we need to show that diagonals *AC* and *BD* have equal lengths. To do this, determine each of their lengths using the distance formula $d^2 = (x_2 - x_1)^2 + (y_2 - y_1)^2$:

$$AC^2 = (x_3 - x_1)^2 + (y_3 - y_1)^2 = (0 - 4)^2 + (0 - 3)^2 = 16 + 9 = 25$$

$$AC = 5$$

$$BD^2 = (x_4 - x_2)^2 + (y_4 - y_2)^2 = (0 - 4)^2 + (3 - 0)^2 = 16 + 9 = 25$$

$$BD = 5$$

Therefore, because *AC* = 5 and *BD* = 5, *AC* is congruent to *BD* and the diagonals of rectangle ABCD are congruent.

Compare this method with proving diagonals are congruent in Chapter 5, Section 5.6., where we used congruent triangles.

• **Example**: For a circle in a coordinate system with its center at C(−3, 2) and a radius of 2, write an equation describing the circle using the distance formula.

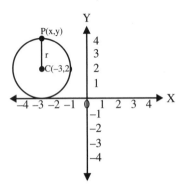

P(x,y) represents any
point on circle C.
r is the radius of circle C.
The center of circle C
is at coordinate (–3,2).

Because P(x, y) represents any point on the circle and C(–3, 2) is the center with the radius r being the distance between C and P, or r = CP, the distance formula, $d^2 = (x_2 - x_1)^2 + (y_2 - y_1)^2$, becomes:

$r^2 = (x - (-3))^2 + (y - 2)^2$

$r^2 = (x + 3)^2 + (y - 2)^2$

This equation describes the circle.

• Note that the equation for a ***circle*** with radius r located at the origin of a coordinate system can be written in the form: $x^2 + y^2 = r^2$, where $r > 0$. For a circle whose center is located at a point (x = p, y = q), or (p, q), other than the origin, the equation becomes: $(x - p)^2 + (y - q)^2 = r^2$.

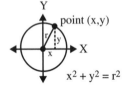

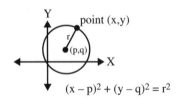

Equation of a circle theorem: An equation describing a circle with radius r whose center is located at a point (x = p, y = q), or (p, q), is:

$(x - p)^2 + (y - q)^2 = r^2$

When the circle is at the origin (0, 0), the equation reduces to:

$x^2 + y^2 = r^2$, where $r > 0$

9.3 Midpoint Formula

• In Chapter 2 *Points, Lines, Planes, and Angles*, Section 2.2., we learned that the ***midpoint of a line segment*** is halfway between each endpoint and is therefore equidistant between the endpoints and divides the segment into two *congruent segments*. Any straight line segment has one and only one midpoint. The midpoint M of a line segment bisects the segment such that in the figure: $AM \cong BM$.

Point M bisects *AB*.

The *midpoint of a segment can be obtained from taking the average of its endpoints*. (Remember, to find the average of n numbers, divide the *sum* of the numbers by n, which is the number of numbers. See *Master Math: Basic Math and Pre-Algebra* for a definition of *average*.)

• We can determine the *coordinate of the midpoint of a line segment on a one-dimensional coordinate system* by describing our line segment using equations.

$AM \cong BM$,
$AM = x - x_1$ and $BM = x_2 - x$.

In the figure, point M is the midpoint of *AB* and $x_2 > x > x_1$ such that $AM \cong BM$. Because the distance between points is the absolute value of their difference, as well as the length of the segment between them, we can describe *AM* and *BM* as:

$AM = x - x_1$ and $BM = x_2 - x$

If we substitute into $AM = BM$ and solve for x:

$$x - x_1 = x_2 - x$$
$$2x = x_1 + x_2$$
$$x = (x_1 + x_2)/2$$

This is the coordinate of the midpoint of *AB* and is written:
$M(x) = (x_1 + x_2)/2$, which describes the midpoint as the average of its endpoints. If this segment was vertical and parallel to the Y-axis, the midpoint would be described as: $M(y) = (y_1 + y_2)/2$

• To determine the *coordinate of the midpoint of a line segment (AB) on a two-dimensional coordinate system* that is not parallel to either axis, we can extend horizontal and vertical lines so that P is the midpoint of *AC* and Q is the midpoint of *BC* (depicted below).

Points M and Q both correspond to a midpoint x-coordinate of:

$$x = (x_1 + x_2)/2$$

Points M and P both correspond to a midpoint y-coordinate of:

$$y = (y_1 + y_2)/2$$

Therefore, the coordinate of the **midpoint** between the two points in a *two-dimensional coordinate system* is:

$$M(x, y) = ((x_1 + x_2)/2, (y_1 + y_2)/2)$$

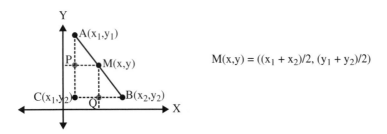

$$M(x,y) = ((x_1 + x_2)/2, (y_1 + y_2)/2)$$

Midpoint formula theorem: The coordinates of the midpoint between the two points (x_1, y_1) and (x_2, y_2) are:

$$M(x, y) = \left(\frac{x_1 + x_2}{2}, \frac{y_1 + y_2}{2} \right)$$

• **Example**: A wildlife refuge has been carefully mapped out using a coordinate system so that specific locations can be pinpointed. It has been determined that an additional ranger station, R_m, is needed between two existing ones, R_1 and R_2. What would be the exact midpoint between the two ranger stations if they are at coordinates $R_1(2, 6)$ and $R_2(10, -3)$. After you determine the coordinate of the midpoint between $R_1(2, 6)$ and $R_2(10, -3)$, use the distance formula to check the result.

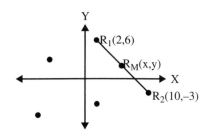

Ranger station $R_1(2,6) = (x_1,y_1)$
New mid-station $R_M(x,y) = (x,y)$
Ranger station $R_2(10,-3) = (x_2,y_2)$

To find the midpoint of segment R_1R_2 we can use the midpoint formula

$M(x, y) = ((x_1 + x_2)/2, (y_1 + y_2)/2)$

and substitute $R_1(2, 6) = (x_1, y_1)$ and $R_2(10, -3) = (x_2, y_2)$:

$M(x, y) = ((2 + 10)/2, (6 + -3)/2) = (12/2, 3/2) = (6, 1.5)$

The midpoint coordinate of segment R_1R_2 is $M(x, y) = (6, 1.5)$.

To check that $M(x, y) = (6, 1.5)$ is the midpoint of R_1R_2 we can use the distance formula $d^2 = (x_2 - x_1)^2 + (y_2 - y_1)^2$ to verify that $R_1M \cong MR_2$.

First consider R_1M, where

$R_1(2, 6) = (x_1, y_1)$ and $M(x, y) = (6, 1.5)$:

$d^2 = (x - x_1)^2 + (y - y_1)^2$

$\quad = (6 - 2)^2 + (1.5 - 6)^2 = (4)^2 + (-4.5)^2 = 16 + 20.25 = 36.25$

$R_1M = d = \sqrt{36.25}$

Next consider MR_2, where

$R_2(10, -3) = (x_2, y_2)$ and $M(x, y) = (6, 1.5)$:

$d^2 = (x_2 - x)^2 + (y_2 - y)^2$

$\quad = (10 - 6)^2 + (-3 - 1.5)^2 = (4)^2 + (-4.5)^2 = 16 + 20.25 = 36.25$

$MR_2 = d = \sqrt{36.25}$

Therefore, because $R_1M \cong MR_2$, the point $M(x, y) = (6, 1.5)$ is the midpoint of R_1R_2 and is where we want our new ranger station.

• **Example**: Prove that the segment connecting the midpoints of two sides of a triangle is parallel to the third side and has a length that is one-half of the length of the third side.

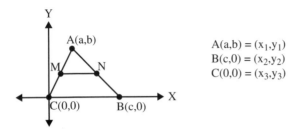

Given: $\triangle ABC$ in which MN connects midpoints of AC and AB.

Prove: $MN \parallel BC$ and $MN = (1/2)BC$.

Proof:

Set $\triangle ABC$ into a coordinate system with one vertex C at the origin.

Determine the coordinates of M and N using the midpoint formula:

$M(x, y) = ((x_1 + x_3)/2, (y_1 + y_3)/2) = ((a + 0)/2, (b + 0)/2) = (a/2, b/2)$

$M(x, y) = M(a/2, b/2)$

$N(x, y) = ((x_1 + x_2)/2, (y_1 + y_2)/2) = ((a + c)/2, (b + 0)/2)$
$= (a/2 + c/2, b/2)$

$N(x, y) = N(a/2 + c/2, b/2)$

Show that $MN \parallel BC$ by comparing their slopes using the slope equation, (which we learn in the next section): Slope $= (y_2 - y_1)/(x_2 - x_1)$, or (change in y) /(change in x).

Slope of $MN = (b/2 - b/2)/(a/2 + c/2 - a/2) = 0/(c/2) = 0$

Slope of $BC = (0 - 0)/(c - 0) = 0/c = 0$

The slopes of MN and BC are both zero, and because slopes of parallel lines are the same, segment MN is parallel to BC, or $MN \parallel BC$.

To demonstrate that $MN = (1/2)BC$, compare the length of segments MN and BC by subtracting x-coordinates and subtracting y-coordinates:

$MN = N(a/2 + c/2, b/2) - M(a/2, b/2) = a/2 + c/2 - a/2 = c/2$, (where $b/2 - b/2 = 0$)

$BC = B(c, 0) - C(0, 0) = c - 0 = c$, (where $0 - 0 = 0$)

Because $MN = c/2$ and $BC = c$, then by substituting $c = BC$ into $MN = c/2$ the result is: $MN = (1/2)BC$

Therefore, we have proved that $MN \parallel BC$ and $MN = (1/2)BC$.

9.4 Slope of a Line Including Parallel and Perpendicular Lines

• The slope of a non-vertical line measures the line's steepness and direction upward, downward, or horizontally. Slope is sometimes referred to by other words such as grade, as in the grade of a road. The letter generally used to symbolize slope is m.

• The *slope of a line on a coordinate system* can be determined by choosing a segment of the line between two convenient points on the line and calculating the change in the vertical y-direction divided by the change in the horizontal x-direction.

Slope = change in y ÷ change in x

The change in the y-direction is determined by subtracting the corresponding Y-axis value at the first y-point from the corresponding Y-axis value at the second y-point: Change in y = $y_2 - y_1$.

The change in the x-direction is determined by subtracting the corresponding X-axis value at the first x-point from the corresponding X-axis value at the second x-point: Change in x = $x_2 - x_1$.

The *equation for the slope m of a line* that contains points $A(x_1, y_1)$ and $B(x_2, y_2)$ is:

Slope m = $(y_2 - y_1) / (x_2 - x_1)$

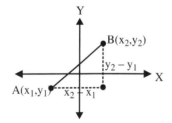

Slope is:

$$\frac{\text{change in y}}{\text{change in x}}$$

or

$$\frac{y_2 - y_1}{x_2 - x_1}$$

• The slope does not depend on the order of the points chosen. For example, in the previous figure if $(x_1, y_1) = (-3, -1)$ and $(x_2, y_2) = (3, 3)$:

Slope m = $(y_2 - y_1) / (x_2 - x_1) = (3 - (-1)) / (3 - (-3)) = 4/6 = 2/3$

Slope m = $(y_1 - y_2) / (x_1 - x_2) = (-1 - 3) / (-3 - 3) = -4/-6 = 4/6 = 2/3$

(A negative number divided by a negative number equals a positive number.)

Therefore, we can write: $(y_2 - y_1) / (x_2 - x_1) = (y_1 - y_2) / (x_1 - x_2)$.

• The slope is sometimes described as the *rise* over the *run*, or slope = rise/run = (change in y) / (change in x). Certain airplanes are capable of steep takeoffs. The rise/run or slope of the airplane in the figure is 3/5. The rise/run or slope of the submarine is –1/5.

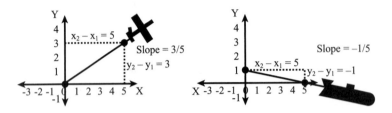

• The relationship between the slope m = $(y_2 - y_1)/(x_2 - x_1)$ and the line:

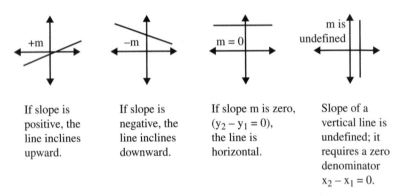

| If slope is positive, the line inclines upward. | If slope is negative, the line inclines downward. | If slope m is zero, $(y_2 - y_1 = 0)$, the line is horizontal. | Slope of a vertical line is undefined; it requires a zero denominator $x_2 - x_1 = 0$. |

• We can determine the slope of the line in the graph below as follows:

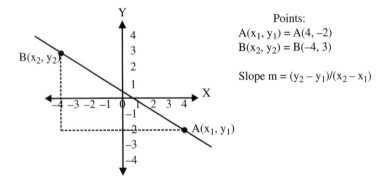

Points:
$A(x_1, y_1) = A(4, -2)$
$B(x_2, y_2) = B(-4, 3)$

Slope m = $(y_2 - y_1)/(x_2 - x_1)$

Choose two points on the line:

$A(x_1, y_1) = A(4, -2)$ and $B(x_2, y_2) = B(-4, 3)$

Calculate slope m using the equation: Slope m $= (y_2 - y_1) / (x_2 - x_1)$

Slope m $= (3 - (-2)) / (-4 - 4) = 5/-8 = -5/8$

Therefore, the slope of the line is $-5/8$. The negative sign indicates that the slope inclines downward.

• Note that the greater the absolute value of the slope, the steeper will be the line; either a steep incline for a large positive slope value or a steep decline for a large negative slope value.

• **Example**: Sketch lines with slopes m $= 4$ and m $= 1/4$. (Remember, $4 = 4/1$.)

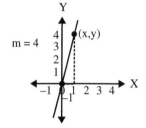
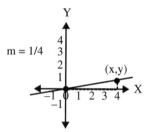

m = (change in y/change in x) = 4/1.
To sketch: Select point (0,0);
move +1 along X, then +4 along Y.
Plot point (x,y). Then draw a
line through (0,0) and (x,y).

m = (change in y/change in x) = 1/4.
To sketch: Select point (0,0);
move +4 along X, then +1 along Y.
Plot point (x,y). Then draw a
line through (0,0) and (x,y).

Note: To sketch a graph if you are given the slope and one point, begin at the known point and move in horizontal-X and vertical-Y directions: right or left and up or down according to the sign of the slope and the change in y and change in x values.

Slopes of Parallel and Perpendicular Lines

• Non-vertical *parallel lines* have the same slope.

Parallel line slope theorem: If two non-vertical lines are parallel, they have the same slope.

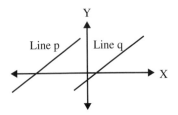

If line p ∥ line q, then $m_p = m_q$.
If $m_p = m_q$, then line p ∥ line q.

Converse of parallel line slope theorem: If two lines have the same slope, they are non-vertical parallel lines.

• How to prove *parallel line slope theorem*: If line p and line q (in the figure) are parallel, they have the same slope.

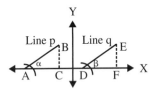

If line p ∥ line q, then $m_p = m_q$.
If $m_p = m_q$, then line p ∥ line q.

To prove this theorem, we draw two congruent segments *AC* and *DF* along the X-axis. At point C and point F, draw segments *BC* and *EF* parallel to the Y-axis and therefore perpendicular to X-axis and perpendicular to respective segments *AC* and *DF* resulting in

mLACB = mLDFE = 90°.

We see that $L\alpha \cong L\beta$ because parallel lines (line p and line q) cut by a transversal (X-axis) form congruent corresponding angles.

By ASA (angle-side-angle) congruence postulate, △ABC ≅ △DEF. Because corresponding parts of congruent triangles are congruent,

$BC \cong EF$.

To show that the slopes are the same, we remember that corresponding sides of congruent triangles are in proportion so that: *BC/EF = AC/DF*, or by cross-multiplication, *(BC)(DF) = (EF)(AC)*, or dividing by *AC* and *DF*, *BC/AC = EF/DF* where *BC/AC* = m_p is the slope of line p and *EF/DF* = m_q is the slope of line q. Therefore, by substituting m_p and m_q into *BC/AC = EF/DF*, then $m_p = m_q$ and the slopes are the same.

• Non-vertical *perpendicular lines* have slopes that are opposite reciprocals and multiply to –1.

Perpendicular line slope theorem: If two non-vertical lines are perpendicular, they have slopes that are opposite reciprocals of one another, ($m_p = -1/m_q$), and the product of their slopes is –1, ($m_p \times m_q = -1$).

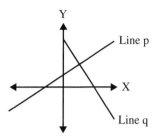

If $p \perp q$, then $m_p = -1/m_q$ or $m_p \times m_q = -1$.
If $m_p = -1/m_q$ or $m_p \times m_q = -1$, then $p \perp q$.

Converse of perpendicular line slope theorem: If the slopes of two lines are opposite reciprocals of one another, or the product of their slopes is –1, then they are non-vertical perpendicular lines.

• How to prove ***perpendicular line slope theorem***: If line p and line q (in the figure) are perpendicular, they have slopes that are opposite reciprocals of one another and the product of their slopes is –1.

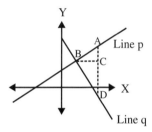

If $p \perp q$, then $m_p = -1/m_q$ or $m_p \times m_q = -1$.
If $m_p = -1/m_q$ or $m_p \times m_q = -1$, then $p \perp q$.

We can prove this theorem by drawing line segments BC and AD, which creates similar right triangles $\triangle ABD$, $\triangle BCD$, and $\triangle ACB$.

In a right triangle, the altitude (BC) to the hypotenuse is the geometric mean between the segments of the hypotenuse (AC and DC). (See Section 4.7. *Similar Right Triangles*.) The proportion is: $AC/BC = BC/DC$, where AC/BC is the slope of line p, or m_p, and $-DC/BC$ is the slope of line q, or m_q. Because $-DC/BC = m_q$, then $BC/DC = -1/m_q$, and $AC/BC = BC/DC$ is equivalent to $m_p = -1/m_q$, which demonstrates that the slopes are opposite reciprocals.

If we rearrange $m_p = -1/m_q$ by cross multiplication, we get:

$m_p \times m_q = -1$, which shows that the product of their slopes is –1.

• Note: Horizontal and vertical lines are always perpendicular so that any two lines where one has a zero slope and the other has an undefined slope are perpendicular.

• **Example:** What is the slope of all lines parallel and all lines perpendicular to PQ if P(–4, 2) and Q(5, 4)?

The slope of PQ is m $= (y_2 - y_1)/(x_2 - x_1) = (4 - 2)/(5 - (-4)) = 2/9$

The slope of all lines parallel to PQ is 2/9.

The slope of all lines perpendicular to PQ is $-1/(2/9) = -9/2$

To check, does $m_p \times m_q = -1$?

$2/9 \times (-9/2) = -18/18 = -1$.

Therefore, the slope of all lines parallel is 2/9, and the slope of all lines perpendicular is –9/2.

9.5 Defining Linear Equations

• A *linear equation* contains only first-degree variables, which indicates there are no exponents or radicals. Examples of linear equations include:

$2x + 3y = 6$; $3x = 2$; $y = 10$; and $x - y = 5$.

If a linear equation is plotted, the result is a straight line.

A *non-linear equation* contains at least one variable with an exponent other than one. Examples of non-linear equations include:

$2x^2 + 3y = 6$; $3x^2 = 2$; $y^4 = 10$; and $x^3 - y = 5$.

If a non-linear equation is plotted, the result is a curve.

• Any *linear equation* can be written in three forms, which are discussed in this section: The *standard form* Ax + By = C, where A, B, and C are constants; the *point-slope form* $(y - y_0) = m(x - x_0)$, where (x_0, y_0) is the known point; and the *slope-intercept form* y = mx + b, where b is the y-intercept. (Also, see *Master Math: Algebra*.)

Standard Form of Equation of a Line (Linear Equation)

• Any linear equation can be written in what is called *standard form of the equation of a line*:

Ax + By = C

Where A, B, and C are real number constants and A and B are not both zero.

In the examples of linear equations listed above:

$2x + 3y = 6,$ $A = 2, B = 3, C = 6$

$3x = 2,$ $A = 3, B = 0, C = 2$

$y = 10,$ $A = 0, B = 1, C = 10$

$x - y = 5,$ $A = 1, B = -1, C = 5$

• *Standard form line equation theorem*: The graph of an equation that is a line can be expressed in the form: $\mathbf{Ax + By = C}$ where A, B, and C are real number constants and A and B are not both zero.

• The standard form $Ax + By = C$ of a linear equation can be *easily graphed* because the points where the line intercepts the X and Y axes can be found by first setting y equal to zero and solving for x, resulting in coordinate $(x, 0)$, which is where the line crosses the X-axis; then setting x equal to zero and solving for y, resulting in coordinate $(0, y)$, which is where the line crosses the Y-axis. The equation can be plotted by drawing a line through the points. (See next section for an example.)

Point-Slope Form of an Equation of a Line (Linear Equation)

• The equation for the *slope of a line*, $m = (y_2 - y_1) / (x_2 - x_1)$, can be algebraically rearranged by multiplying both sides by $(x_2 - x_1)$ as:

$$(y_2 - y_1) = m(x_2 - x_1)$$

This equation is called the ***point-slope form of an equation of a line***, and is useful in finding the equation of a line if the coordinates of one point on the line and the slope of the line are known.

The ***point-slope form of an equation of a line*** is often written:

$$(y - y_0) = m(x - x_0)$$

Where (x_0, y_0) represents the known point on the line and m is the slope.

• *Point-slope line equation theorem*: The point-slope form of an equation of a line passing through known point (x_0, y_0) and having slope m is:
$(\mathbf{y - y_0}) = \mathbf{m(x - x_0)}$

• **Example:** Find the equation of the line $(y - y_0) = m(x - x_0)$ if:

(a) the slope is -3 and one point is $(-1, 4)$;

(b) two known points are $(-2, 3)$ and $(5, 4)$;

(c) one point is $(1, 1)$ and the line is perpendicular to $y = 3x - 4$;

(d) one point is $(1, 1)$ and the line is parallel to $y = 3x - 4$; and

(e) we transform the equation found in (a) to standard form.

(a) $(y - y_0) = m(x - x_0)$, or $(y - 4) = -3(x - (-1))$, or $(y - 4) = -3(x + 1)$.

(b) First determine the slope $m = (y_2 - y_1) / (x_2 - x_1)$, or
$m = (4 - 3)/(5 - (-2)) = 1/7$. Determine the line equation,
$(y - y_0) = m(x - x_0)$, using either known point:

$(y - 4) = (1/7)(x - 5)$, or $(y - 3) = (1/7)(x + 2)$.

(c) Because the line is perpendicular to $y = 3x - 4$, which has a slope
of 3, the slope of a perpendicular line will be $-1/3$. The line in
point-slope form, $(y - y_0) = m(x - x_0)$, with point $(1, 1)$ is:

$(y - 1) = (-1/3)(x - 1)$.

(d) Because the line is parallel to $y = 3x - 4$, which has a slope of 3, the
slope of a parallel line will also be 3. The line in point-slope form,
$(y - y_0) = m(x - x_0)$, with point $(1, 1)$ is: $(y - 1) = (3)(x - 1)$.

(e) The standard form of an equation for a line is $Ax + By = C$, and the
equation in (a) is: $y - 4 = -3(x + 1)$

Rearrange the equation in (a) to $Ax + By = C$ using algebra:

$y - 4 = -3x - 3$

$3x + y = 4 - 3$

$3x + y = 1$, which is in standard form with $A = 3$, $B = 1$, and $C = 1$.

Slope-Intercept Form of the Equation of a Line (Linear Equation)

• In the *point-slope equation*, $(y - y_0) = m(x - x_0)$, if known point
(x_0, y_0) falls on the Y-axis, (where the line crosses the Y-axis),
then $x_0 = 0$ and the point becomes $(0, y_0)$.

Substituting $(0, y_0)$ into the *point-slope equation*, $(y - y_0) = m(x - x_0)$,
results in:

$y - y_0 = mx$, or $y = mx + y_0$

This equation is commonly written with the **y-intercept $y_0 = b$**, where
the line crosses the Y-axis and is called the **slope-intercept form of the
equation of a line**.

$y = mx + b$

Where $b = y_0$ is the y-intercept of the line and m is the slope.

• **Slope-intercept line equation theorem**: The equation of a line having
slope m and y-intercept b is: $y = mx + b$

- The *slope-intercept form of the equation of a line* can be used to determine the slope m and y-intercept b when the equation for the line is known.

- **Example:** If the equation for a line is $2x + 6y = 12$, determine the slope m and y-intercept b. Then sketch the line.

Simplify the equation and algebraically rearrange it into the *slope-intercept form* $y = mx + b$:

$2x + 6y = 12$

$x + 3y = 6$

$3y = -x + 6$

$y = (-x + 6)/3$

$y = (-1/3)x + 2$

The slope of the line is $m = -1/3$ and the line crosses the Y-axis at 2.

We can sketch the line by identifying the y-intercept, 2, with a point. Then use the slope, $m = -1/3 =$ (change in y)/(change in x), to move either left three units and up one unit, or right three units and down one unit.

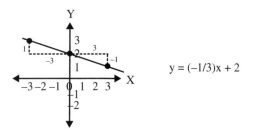

$y = (-1/3)x + 2$

- The equation for the slope of a line, $m = (y_2 - y_1)/(x_2 - x_1)$, can be used to find the equation for a line if two points (x_1, y_1) and (x_2, y_2) are known. Because $m = (y_2 - y_1)/(x_2 - x_1)$:

$$\frac{(y - y_1)}{(x - x_1)} = \frac{(y_2 - y_1)}{(x_2 - x_1)}$$

This equation is the **two-point form of the equation** of a line, and it can be used to determine the slope of the line. Then, the known slope value along with one known point can be substituted into the *point-slope equation*: $(y - y_0) = m(x - x_0)$.

9.6 Graphing Linear Equations

• A *line* can be defined by either a two-axis or a three-axis coordinate system, such that each point on the line corresponds to a coordinate on each axis. A line can be described by a minimum of two points that correspond to positions on the X and Y axes, or by a linear equation that contains variables x and y that correspond to points on the line.

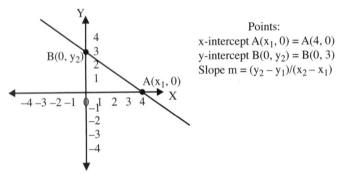

Points:
x-intercept $A(x_1, 0) = A(4, 0)$
y-intercept $B(0, y_2) = B(0, 3)$
Slope $m = (y_2 - y_1)/(x_2 - x_1)$

x-intercept of a line is the point where it crosses the X-axis and y = 0: (x, 0)

y-intercept of a line is the point where it crosses the Y-axis and x = 0: (0, y)

• One method in which an equation for a line can be graphed is as follows:

Simplify the equation by combining like terms and arrange it into standard form: $Ax + By = C$.

Determine the points where the line intercepts the X and Y axes: First set y equal to zero, and solve for the x-intercept, resulting in coordinate (x, 0), which is where the line crosses the X-axis. Then set x equal to zero, and solve for the y-intercept, resulting in coordinate (0, y), which is where the line crosses the Y-axis.

Graph the line by plotting the two points at the intercepts and draw a line through the points.

Check that the result is correct by choosing an x value, substituting it into the original equation, and solving for the corresponding y value. This (x, y) coordinate should fall on the line.

• **Example:** Graph the equation $4x - 8x - 8 = 2y + 6y + 8$.

Simplify and arrange into standard form, $Ax + By = C$:

$-4x - 8 = 8y + 8$, or $4x + 8y = -16$, or $x + 2y = -4$

Determine the y-intercept by setting x = 0: $2y = -4$, or

$y = -4/2 = -2$, or $y = -2$

The y-intercept is (0, –2)

Determine the x-intercept by setting y = 0: x = –4

The x-intercept is (–4, 0)

Graph the line by plotting the two axis-intercept points:

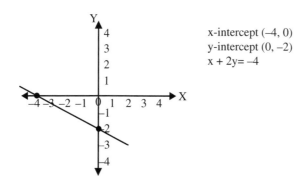

x-intercept (–4, 0)
y-intercept (0, –2)
x + 2y= –4

- Another method in which an equation for a line can be graphed includes:

Simplify the equation by combining like terms.

Choose at least two values for x and solve the equation for corresponding y values resulting in coordinates (x, y).

Check that the two coordinates satisfy the equation by substituting them into the original equation.

Graph the line by plotting the two points on the coordinate system and draw a line through the points.

- **Example**: Graph the equation 4y = 4x + 8.

    Simplify by dividing by 4: y = x + 2

    Choose x = 1 and solve for y: y = 1 + 2 = 3

    The first coordinate is (1, 3). Check: 3 = 1 + 2.

    Choose x = –1 and solve for y: y = –1 + 2 = 1

    The second coordinate is (–1, 1). Check: 1 = –1 + 2.

    Choose x = –2 and solve for y: y = –2 + 2 = 0

    The third coordinate is (–2, 0). Check: 0 = –2 + 2.

Graph the line by plotting the points:

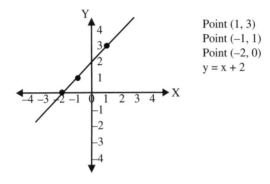

Point (1, 3)
Point (–1, 1)
Point (–2, 0)
y = x + 2

• **Example:** (a) Graph x = 2. (b) Graph y = 2.

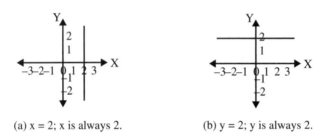

(a) x = 2; x is always 2.   (b) y = 2; y is always 2.

Locus

• *Locus* is also used in analytic, or coordinate, geometry. The locus is the set of points (or graph) of an equation that is given by the coordinates (x, y), or (x, y, z) for three dimensions, that satisfy the equation. In other words, the locus, or graph, of an equation containing two variables, such as x and y, is the curve or line that contains all of the points, and no others, whose coordinates satisfy the equation. (To satisfy an equation, the coordinate values of the x and y variables, when substituted into the equation, must make both sides of the equation equal). For example, the locus of the equation x = 3 is the set of points and only those points that lie three units to the right of the Y-axis and parallel to it. The equation is x = 3, and the locus is the line representing that equation.

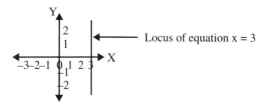

Locus of equation x = 3

9.7 Chapter 9 Summary and Highlights

• A number line is a *one-dimensional coordinate system*. If a number is represented by a point on the number line, the number is called the *coordinate* of that point. A number can be represented in one, two, or three dimensions. A *point* on a planar rectangular coordinate system is represented in two dimensions, x and y. To define the position of a point on a two-axis planar coordinate system, the numbers on each axis that the point corresponds to are the coordinates of the point. Each point on an X-Y coordinate system corresponds to a unique *ordered pair* of real numbers (x, y), which are called the **coordinates of the point** written as: (X-axis-number, Y-axis-number).

• In a coordinate system, if the coordinates of two points are (x_1, y_1) and (x_2, y_2), then the distance d between the two points is given by the **distance formula**: $d^2 = (x_2 - x_1)^2 + (y_2 - y_1)^2$.

• In a coordinate system, the coordinates of the midpoint between the two points (x_1, y_1) and (x_2, y_2) is given by the **midpoint formula**:

$$M(x, y) = ((x_1 + x_2)/2, (y_1 + y_2)/2).$$

• The *slope of a line on a coordinate system* is the change in the vertical y-direction divided by the change in the horizontal x-direction. The **equation for the slope m of a line** that contains points (x_1, y_1) and (x_2, y_2) is:

Slope $m = (y_2 - y_1) / (x_2 - x_1)$.

• Non-vertical **parallel lines** have the same slope, $(m_p = m_q)$. Non-vertical **perpendicular lines** have slopes that are opposite reciprocals and multiply to -1, $(m_p = -1/m_q)$ and $(m_p \times m_q = -1)$.

• Any linear equation can be written in (or graphed as) what is called the *standard form of the equation of a line*: $Ax + By = C$, where A, B, and C are real number constants, and A and B are not both zero.

• The *point-slope form of an equation of a line* passing through known point (x_0, y_0) and having slope m is:

$$(y - y_0) = m(x - x_0).$$

• The equation for a line having slope m and y-intercept b is called the *slope-intercept form of the equation of a line*:

$y = mx + b.$

• A *line* can be defined by a coordinate system, such that each point on the line corresponds to a coordinate on each axis. A line can be described by a minimum of two points that correspond to positions on the X and Y axes, or by a linear equation that contains variables x and y that correspond to points on the line. Lines can be graphed by determining the points where the line intercepts the X and Y axes or by choosing values for x and solving the equation of the line for corresponding y values.

Index